Target
Get back on track

GRADE 5

AQA GCSE (9–1)
Combined Science: Trilogy

Mark Grinsell, Pauline Lowrie,
Ali Mclachlan, Katherine Pate,
Frank Sochacki, Jason Welch

Published by Pearson Education Limited, 80 Strand, London, WC2R ORL.

www.pearsonschoolsandfecolleges.co.uk

Text and illustrations © Pearson Education Ltd 2017
Typeset and illustrated by Tech-Set Ltd, Gateshead
Produced by Haremi

The rights of Mark Grinsell, Pauline Lowrie, Ali Mclachlan, Katherine Pate, Frank Sochacki, Jason Welch to be identified as authors of this work have been asserted by them in accordance with the Copyright, Designs and Patents Act 1988.

First published 2017

20 19 18 17
10 9 8 7 6 5 4 3 2

British Library Cataloguing in Publication Data
A catalogue record for this book is available from the British Library

ISBN 978 0435 18901 3

Acknowledgements
The author and publisher would like to thank the following individuals and organisations for permission to reproduce photographs:

Science Photo Library Ltd: Dr Gopal Murti 46; **Alamy Stock Photo**: Science History Images 48; PjrStudio 137, 141

Note from the publisher
Pearson has robust editorial processes, including answer and fact checks, to ensure the accuracy of the content in this publication, and every effort is made to ensure this publication is free of errors. We are, however, only human, and occasionally errors do occur. Pearson is not liable for any misunderstandings that arise as a result of errors in this publication, but it is our priority to ensure that the content is accurate. If you spot an error, please do contact us at resourcescorrections@pearson.com so we can make sure it is corrected.

 This workbook has been developed using the Pearson Progression Map and Scale for Science.

To find out more about the Progression Scale for Science and to see how it relates to indicative GCSE (9–1) grades go to www.pearsonschools.co.uk/ProgressionServices

Helping you to formulate grade predictions, apply interventions and track progress.

Any reference to indicative grades in the Pearson Target Workbooks and Pearson Progression Services is not to be used as an accurate indicator of how a student will be awarded a grade for their GCSE exams.

You have told us that mapping the Steps from the Pearson Progression Maps to indicative grades will make it simpler for you to accumulate the evidence to formulate your own grade predictions, apply any interventions and track student progress. We're really excited about this work and its potential for helping teachers and students. It is, however, important to understand that this mapping is for guidance only to support teachers' own predictions of progress and is not an accurate predictor of grades.

Our Pearson Progression Scale is criterion referenced. If a student can perform a task or demonstrate a skill, we say they are working at a certain Step according to the criteria. Teachers can mark assessments and issue results with reference to these criteria which do not depend on the wider cohort in any given year. For GCSE exams however, all Awarding Organisations set the grade boundaries with reference to the strength of the cohort in any given year. For more information about how this works please visit: https://www.gov.uk/government/news/setting-standards-for-new-gcses-in-2017

Contents

(1) Diffusion, osmosis and active transport

This unit will help you to understand how different substances pass through the cell membrane to get into cells or to leave cells.

This unit will help you to recognise and describe diffusion, osmosis and active transport.

In the exam, you will be asked to tackle questions such as the one below.

Exam-style question

1 All living cells must exchange substances with their environment. To enter or leave cells these substances must cross the cell membrane. Substances can cross the cell membrane in three ways:

diffusion, osmosis or **active transport**.

1.1 Complete the table below to describe how these substances enter or leave the cell.

Substance entering or leaving the cell	How the substance crosses the cell membrane
Minerals entering a plant root from the soil	
Oxygen entering a liver cell	
Water entering a plant root from the soil	
Glucose taken up from the small intestine	

(4 marks)

You will already have done some work on this topic. Before starting the **skills boosts**, rate your confidence in understanding the three ways in which substances cross the cell membrane. Colour in the bars.

① How do I know when diffusion will occur?	② How do I know when osmosis will occur?	③ How do I know when active transport will occur?

The movement of liquids and gases into and out of cells can be described by **diffusion**, **osmosis** and **active transport**. Each process depends upon differences in concentration to influence the movement of molecules.

Molecules are always moving. They move from an area of high concentration to an area of low concentration. This evens out the concentration and is called **diffusion**. Diffusion is a **passive** process, which means it does **not** require energy to make it happen.

(1) Draw an arrow 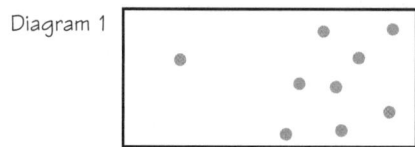 on diagram 1 to show where the molecules will move.

Diagram 1

Osmosis is a form of diffusion to do with movement of water from high concentration to low concentration. It is also **passive** (requiring no energy), and it happens across a **partially permeable** membrane such as a cell membrane. This membrane works like a sieve allowing small molecules (like water) to pass through but not larger molecules (like glucose).

(2) In diagram 2, which molecules can pass through the membrane? 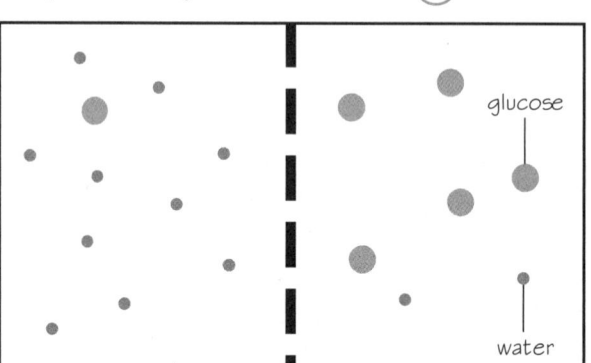 ....

Diagram 2

glucose

Look at the size of the molecules and the size of the holes in the membrane.

water

membrane —

Remember In osmosis only the water molecules move through the membrane. A dilute solution has more water molecules so they will move through the membrane to the concentrated solution, where there are fewer. This is the only way to make the two solutions the same concentration.

(3) In the diagram in (2), what will happen to the concentration of glucose molecules on the right of the membrane as water enters that part of the solution?

Circle (A) the best answer.

Water molecules will move through the membrane until the concentration of glucose molecules is the same on both sides of the membrane.

| stays the same | concentration rises | concentration falls |

Sometimes molecules need to move from a low concentration to a high concentration. This is known as **active transport** because this process requires **energy** from respiration.

(4) Which of these diagrams illustrates active transport? Circle (A) the correct letter.

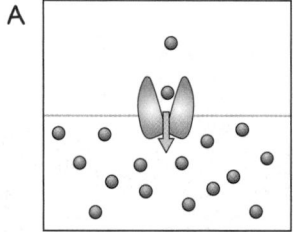
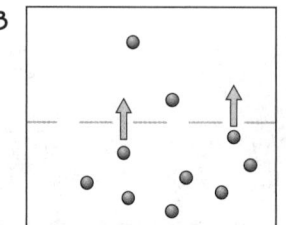
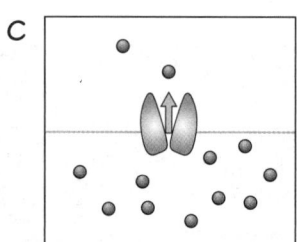

Skills boost

① How do I know when diffusion will occur?

Remember that diffusion is the movement of molecules from where they are more concentrated to where they are less concentrated. This means the molecules become evenly spread out.

① Which diagram represents diffusion?
Circle Ⓐ the correct letter.

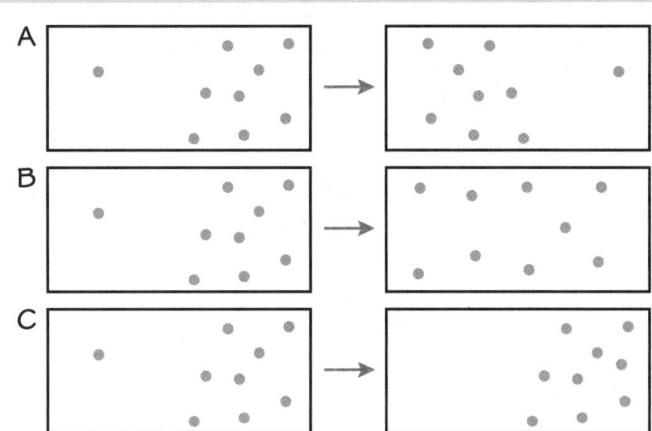

Oxygen and carbon dioxide can easily diffuse across a cell membrane.

Oxygen is used inside cells during respiration. Therefore, oxygen will be at a low concentration inside cells.

Carbon dioxide is made during respiration. Therefore, the concentration of carbon dioxide inside cells is often high.

Answer ✎ this exam-style question.

Biology

Exam-style question

1 The diagram shows molecules of oxygen and carbon dioxide in solution.

Both types of molecules can pass through the cell membrane.

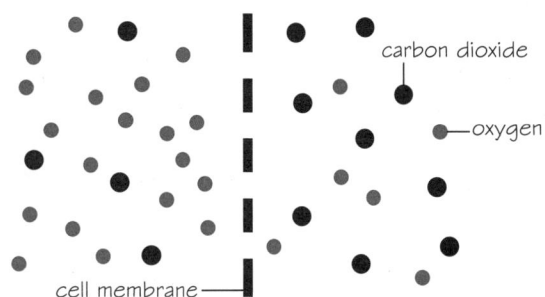

1.1 Which side of the diagram shows a higher concentration of oxygen?

.... (1 mark)

1.2 Which side of the diagram shows the inside of the cell?   **Remember** This is where the carbon dioxide is made during respiration.

.... (1 mark)

1.3 In which direction(s) will the molecules diffuse? Tick **one** box.

Both to the right ☐

Both to the left ☐

Oxygen to the left and carbon dioxide to the right ☐

Carbon dioxide to the left and oxygen to the right ☐

Remember Diffusion occurs *down* the concentration gradient.

(1 mark)

1.4 Explain why oxygen will diffuse in the direction you have stated.   'Explain' means that you must say *why* it happens.

....

....

.... (2 marks)

Oxygen and carbon dioxide will always diffuse into and out of cells down their concentration gradients.

Biology Unit 1 Diffusion, osmosis and active transport   **3**

How do I know when osmosis will occur?

Osmosis is a special form of diffusion. Osmosis is the diffusion of water molecules through a partially permeable membrane.
- The water molecules move from a dilute (less concentrated) solution to a more concentrated solution.
- A dilute solution has more water in it than a concentrated solution.
- So water diffuses from where there are more water molecules to where there are fewer water molecules.

① In an osmosis practical the mass of four potato slices was recorded and they were placed in four different solutions for 24 hours. The mass of each one after 24 hours was then recorded again. The results are shown in the table below.

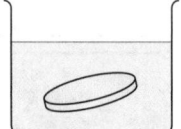 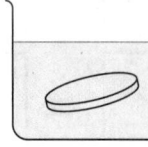 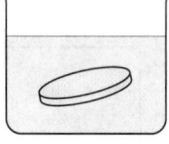 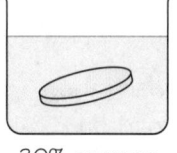

Distilled water   10% sucrose   20% sucrose   30% sucrose

> Any question that asks about osmosis will be about the movement of water into or out of cells.

Solution	Distilled	10% sucrose	20% sucrose	30% sucrose
Initial mass (g)	10	10	10	10
Final mass (g)	11	10	9	8

a Calculate the % change in mass of each of the potato slices.

$$\text{Percentage change} = \frac{\text{actual change}}{\text{original amount}} \times 100$$

	Distilled	10% sucrose	20% sucrose	30% sucrose
Change in mass (g) = final mass − initial mass				$(8 - 10) = -2\,g$
% change in mass $= \dfrac{\text{change in mass}}{\text{initial mass}} \times 100$				$\dfrac{-2}{10} \times 100$ $= -0.2 \times 100$ $= -20\%$

b Which piece gained mass?

> A positive percentage change is an increase in mass. A negative percentage change is a decrease in mass.

....

c Explain the reason for the 10% sucrose solution result.

....

....

d Why is using % change in mass a more useful measurement than change in mass in grams?

....

....

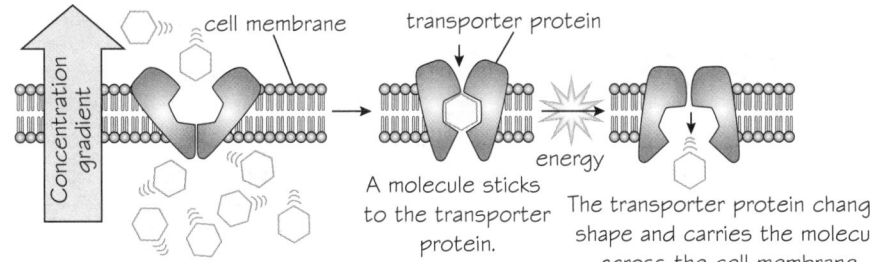

Skills boost

3 How do I know when active transport will occur?

Active transport is different from both diffusion and osmosis. It uses **energy** from respiration in the cell and **transporter proteins** in the membrane to transport molecules across **against** their concentration gradient.

The diagram shows how active transport of a molecule occurs.

cell membrane   transporter protein

Concentration gradient

A molecule sticks to the transporter protein.

energy

The transporter protein changes shape and carries the molecule across the cell membrane.

1 Complete 🖉 the paragraph using information from the diagram.

Active transport uses a .... in the membrane.

A molecule fits into the protein and .... is needed to carry the molecule

across the membrane. The molecule moves .... the concentration gradient.

2 Complete the following table. Place a tick ✓ or a cross ✗ in every box. This will give you a useful summary about diffusion, osmosis and active transport.

Transport process	Molecules move down the concentration gradient?	Uses energy from respiration?	Molecules move against the concentration gradient?
Diffusion			
Osmosis			
Active transport			

Exam-style question

1 The table shows the changes in concentration of glucose along the small intestine after a pasta meal.

All the glucose is absorbed from the small intestine into the blood.

Distance from stomach (cm)	Concentration of glucose (mmol/dm³)
0	0
200	400
400	500
600	200
800	0

1.1 Using information from the table, explain how we know that all the glucose is absorbed.

Look at the data for concentration.

.... **(1 mark)**

1.2 Glucose is absorbed by active transport. Explain how active transport can make the concentration of glucose fall to zero between 600 and 800 cm along the intestine.

You are told this is active transport. How is active transport different from diffusion?

....

.... **(3 marks)**

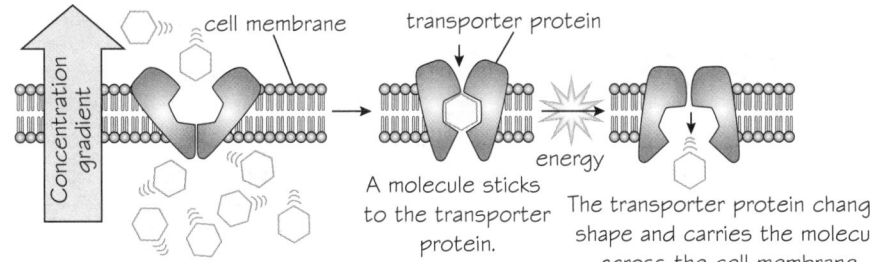

Biology

Sample response

You could be asked to apply your knowledge of the processes by which substances enter and leave living cells.

Look at this exam-style question and the answers given by a student.

Exam-style question

1 The table shows a student's observations on the effect of growing radish seedlings in different types of water.

Water provided	Average height of seedlings after five days (cm)	Observations
Tap water	8.2	Growing well
Sea water	4	All plants have shrivelled and died
Distilled water (contains no minerals)	5.1	Yellowish leaves
Boiled and cooled tap water (contains no oxygen)	7.5	I thought these would die

1.1 Explain the results observed for sea water.

All the plants have shrivelled and died.   **(2 marks)**

1.2 Suggest why the plants did not grow well with distilled water.

There were no mineral ions in the water.   **(2 marks)**

1.3 Before the experiment it was predicted that the seedlings grown in boiled and cooled water would die because they had no oxygen. Suggest how the seedling roots might have gained oxygen.

The oxygen diffused in and was used for respiration.   **(2 marks)**

1 a Why is the response to **1.1** incorrect? ✏   Look at the command word

....

b What process involving substances entering and leaving cells is being tested in **1.1**? ✏   This is about the movement of water into and out of cells.

....

c What would be a better response to **1.1**? ✏   Think about the size of the molecules and the type of membrane.

....

....

2 Describe ✏ how mineral ions are taken in by plants and how they are used.

....

....

Your turn!

Answer the exam-style question using the guided steps below.

Exam-style question

1   All living cells must exchange substances with their environment. To enter or leave cells these substances must cross the cell membrane. Substances can cross the cell membrane in three ways:

diffusion, osmosis or **active transport**.

    1.1   Complete the table below to describe how these substances enter or leave the cell.

Look at the command word 'describe'. This means some detail is required.

Substance entering or leaving cell	How the substance crosses the cell membrane
Minerals entering a plant root from the soil	
Oxygen entering a liver cell	
Water entering a plant root from the soil	
Glucose taken up from the small intestine	

You must draw on your knowledge from other areas of the syllabus. You will need to know that the concentration of minerals in the soil is lower than inside the plant cells.

Oxygen is used inside cells. Therefore, oxygen is at a low concentration inside the liver cell.

Any question that asks about the movement of water into or out of cells will be asking about osmosis.

Glucose is a useful nutrient, therefore all of the glucose in food is taken up into the cells of the small intestine.

(4 marks)

Need more practice?

You could be questioned about a practical you have carried out such as osmosis in potatoes.

Have a go at this exam-style question.

Exam-style question

1   A student investigated the movement of water into and out of potato tissue. This was his method:

   1   Cut five potato sticks that are each 5 cm long.

   2   Calculate the mass of each potato stick and place in separate boiling tubes.

   3   Place each potato stick into sugar solutions at different concentrations.

   4   Leave overnight then recalculate the mass of each potato stick.

   The student's results are shown in the table.

Concentration of sugar solution (mol)	First mass of potato stick (g)	Final mass of potato stick (g)	Change in mass of potato stick (g)	Change in mass of potato stick (%)
1.00	1.85	1.65	−0.2	
0.75	1.70	1.58	−0.12	
0.50	1.80	1.70	−0.10	
0.25	1.96	1.91	−0.05	
0.00	1.76	1.87	0.11	

   1.1   Complete the blank column to improve the validity of the student's results.   **(2 marks)**

   1.2   Explain why the potato chip gained mass at sugar concentration of 0.00.

   ....

   ....

   ....   **(3 marks)**

Boost your grade

To improve your grade, make sure you know these main examples:

- Diffusion of gases in the lungs and diffusion of urea out of the liver.
- Osmosis investigation using potato slices and a range of salt or sugar concentrations.
- Active transport of minerals into plant root cells and of glucose into the cells of the small intestine.

How confident do you feel about each of these **skills?** Colour in the bars.

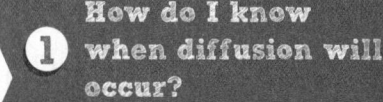

1 How do I know when diffusion will occur?

2 How do I know when osmosis will occur?

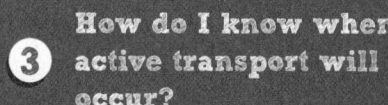

3 How do I know when active transport will occur?

② Enzymes

This unit will help you to understand how enzymes work to speed up biological reactions and what factors affect the speed of enzyme-controlled reactions.

In the exam, you will be expected to tackle questions such as the one below.

Exam-style question

1   A student investigated the effect of pH on the activity of an enzyme that digests starch.

The student recorded how long it took for all the starch to be digested.
The results are shown in the table.

pH	2	5	7	9
Time for complete digestion in seconds		305.0	180.0	280.0
Rate of reaction (s⁻¹)	0	0.33	0.55	0.36

1.1   Describe how changing the pH affects the rate of this reaction.

.... (3 marks)

1.2   Which pH gave the fastest reaction?

.... (1 mark)

1.3   Explain why the reaction was fastest at this pH.

.... (2 marks)

1.4   No time was recorded for pH 2. This is because the starch was not digested.

Explain why the enzyme was unable to digest the starch at pH 2.

.... (2 marks)

You will already have done some work on enzymes. Before starting the **skills boosts**, rate your confidence in each area. Colour in 🖉 the bars.

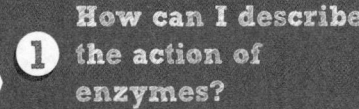

① How can I describe the action of enzymes?

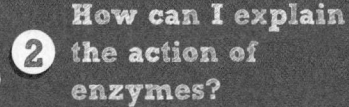

② How can I explain the action of enzymes?

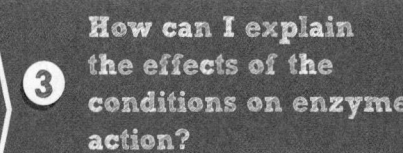

③ How can I explain the effects of the conditions on enzyme action?

Biology

Enzymes are biological **catalysts** that speed up the rate of reactions. They are essential for all the reactions in the body. To understand how enzymes work you need to think about:

- the shapes of the individual molecules

- how their shape is affected by the surrounding conditions.

The substances that enzymes work on are called **substrates**. The substances that are produced are called **products**. Enzymes are always specific, which means that an enzyme will work with only one molecule (its substrate). The shape of the substrate molecule matches the shape of the enzyme. They fit together a bit like two pieces in a jigsaw puzzle.

(**1**) The diagram shows an enzyme and four possible substrates.

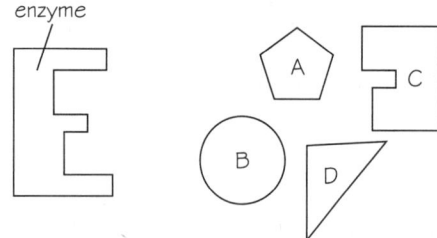

a Circle (A) the substrate molecule that will fit the enzyme.

b Highlight (✏️) the space next to the enzyme where the substrate molecule will fit.

The substrate fits into the **active site** of the enzyme. The way the shapes of the enzyme and the substrate fit together perfectly is described as the **lock-and-key model**.

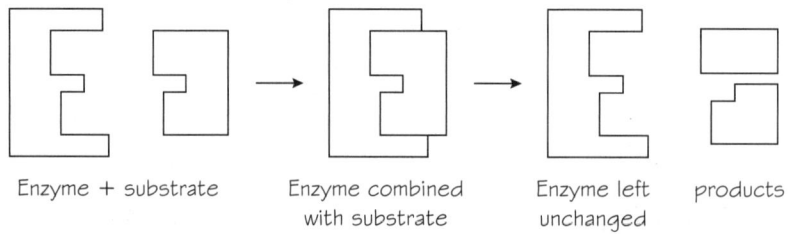

| Enzyme + substrate | Enzyme combined with substrate | Enzyme left unchanged | products |

Remember A catalyst is a substance that speeds up the rate of a reaction without itself being used up.

High temperatures and **extremes of pH** affect the shape of the enzyme's active site.

(**2**) The diagram below shows the shape of the same enzyme after it has been boiled.

Circle (A) the best word from the choice of bold words, to complete these sentences.

High temperatures and extremes of pH change the shape of the **substrate** / **active site**.

This means the substrate will not **fit** / **collide**. We say that the enzyme has become

reformed / **denatured**.

How can I describe the action of enzymes?

To describe the action of an enzyme, you need to state clearly what effect the enzyme has on the reaction. If you are given data, you need to look for a trend and describe it, using values to support your description.

1 The table shows some data about the rate of an enzyme-catalysed reaction.

Temperature (°C)	Rate of reaction (in grams of product formed per minute)
20	7
30	39
40	54
50	32
60	11

a Circle (A) the correct words in these sentences. First you have to look at the effect of the enzyme.

> The enzyme breaks down the **substrate** / **product**. The more **substrate** / **product** formed per minute, the **slower** / **faster** the rate of reaction.

b Use the data in the table to complete the sentences. Then look at the data and describe the trends. Use values to support the description.

> As the temperature increases from ....°C to ....°C, the rate of reaction increases.
>
> The reaction is fastest at ....°C. As the temperature increases to ....°C, the rate of
>
> reaction decreases.

2 A student placed 10 cm³ of a protein suspension into a test tube and added 1 cm³ of a protein-digesting enzyme. The student measured the concentration of amino acids in the tube every 5 minutes. The graph shows the results.

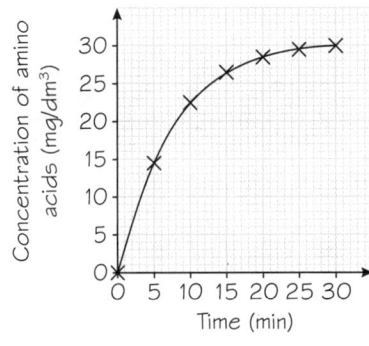

a What is the substrate? Underline (A) it in the question.

b What is the product? Highlight it in the question.

c Describe what is happening in the reaction.

....

....

d What is the shape of the line on the graph? Tick (✓) the box that gives the best description.

Stays level with no change	☐
Rises slowly at first then rises more steeply	☐
Rises steeply at first, then more slowly and then levels off	☐

Look at the shape of the curve. This shows you the trend in the data.

e Describe the trend, using the description you have chosen in **d** and data values from the graph.

Give the values where the shape of the graph changes.

As time increases. ....

....

Biology

② How can I explain the action of enzymes?

For a reaction to take place, the enzyme and the substrate molecules must collide. When the concentration is higher, there are more molecules. When there are more molecules, there are more collisions. When there are more collisions, the rate of reaction is increased.

① Look at this table of results.

Substrate concentration (mol/dm³)	0	0.5	1.0	1.5	2.0
Rate of reaction (mg/s)	0	3	6	9	12

Think carefully about the number of molecules of substrate and how often collisions will occur.

a When is the rate of reaction highest? Circle Ⓐ the substrate concentration in the table.

b Explain 🖉 why the rate of reaction is highest at this point by completing these sentences.

When the substrate concentration is zero ....

....

As the substrate concentration increases ....

....

The highest number of collisions is when ....

....

② The table below shows a range of conditions (A, B, C and D) in which reactions can occur. Condition C is present at the start of the reaction.

	A	B	C	D
Concentration of enzyme	high	low	high	low
Concentration of substrate	low	low	high	high

Remember that the **enzyme** will be left unchanged by the reaction, but the number of **substrate** molecules decreases during the reaction.

Circle Ⓐ the correct letter to answer these questions.

a Which conditions will cause the molecules to collide most often? A B C D

b Which conditions are found at the end of a reaction? A B C D

③ The enzyme amylase converts starch (the substrate) to maltose. The graph shows how the concentration of the product maltose changes during the reaction.

In each question, think about the number of molecules of starch.

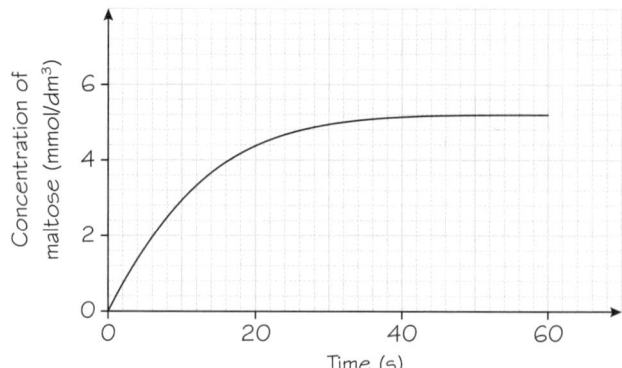

a Explain 🖉 why the reaction has stopped at 60 s.

....

b Explain 🖉 why the reaction slows down between 20 s and 40 s.

....

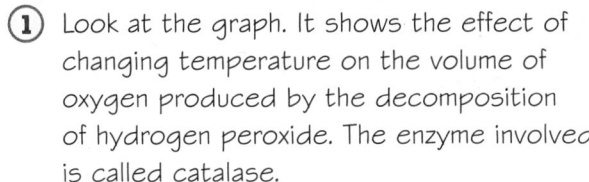

How can I explain the effects of the conditions on enzyme action?

At higher temperatures, molecules have more energy and are more likely to collide. Enzymes have an optimum temperature and pH at which they work most quickly. Extremes of pH or high temperatures can affect the shape of the active site, so that the enzyme becomes denatured and no longer catalyses the reaction.

① Look at the graph. It shows the effect of changing temperature on the volume of oxygen produced by the decomposition of hydrogen peroxide. The enzyme involved is called catalase.

First check you understand the reaction.

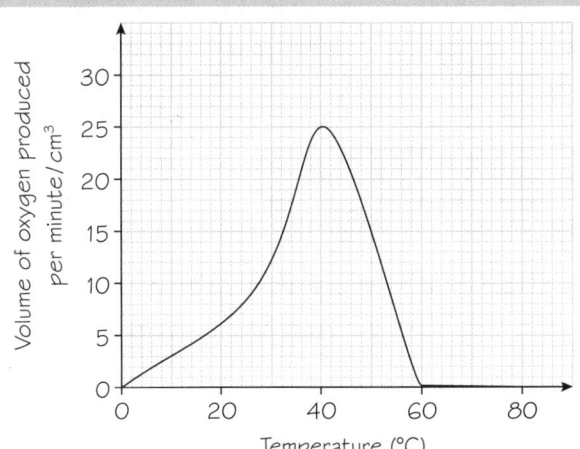

a In the question above, circle Ⓐ the substrate and underline Ⓐ the product.

b Circle Ⓐ the correct words.

> At low temperature (5 °C) the molecules have **lots of** / **very little**
>
> energy, so the molecules will be moving **very quickly** / **very slowly**.
>
> At this temperature the molecules will collide and react
>
> **very often** / **not very often**.

To explain the shape of the graph you need to think about the movement of the molecules.

c Complete ✎ the sentences.

Molecules are always moving. The higher the temperature, the faster they move.

• At higher temperatures (40 °C) molecules have .... energy.

• At 40 °C molecules will collide and react .... than at 5 °C.

d At what temperature is the volume of oxygen collected highest? Circle Ⓐ the best answer.

The **optimum** condition is where the enzyme works most quickly.

> 0 °C   10 °C   20 °C   30 °C   40 °C   50 °C   60 °C   70 °C

e What happens to the volume of oxygen collected as the temperature is increased above 60 °C? ✎

....

f Explain ✎ what has happened to the enzyme above 60 °C.   What happens to the shape of the active site if the temperature rises too high?

....

....

....

Sample response

A good explanation about enzyme activity should:
- include key terms such as 'active site'
- refer to values from a graph or table if you are asked to interpret data
- explain rates of reaction in terms of numbers of molecules and collisions
- explain the effect of pH and temperature in terms of enzyme shape.

Look at this exam-style question and the answers given by a student.

Exam-style question

1   A student investigated the digestion of fat by lipase.

   1.1   Name the substrate and the enzyme.

   The substrate is fat and the enzyme is lipase.

   .... (2 marks)

   1.2   The student expected the reaction to be most rapid at the start of the experiment.

   Explain why the reaction should be fastest at the start.

   This is when the concentration of fat molecules is highest.

   .... (3 marks)

   The student altered the temperature and measured the rate of reaction at each temperature. The student plotted the results on the graph.

   1.3   What is the optimum temperature?

   40 °C

   .... (1 mark)

   1.4   Explain why the reaction occurs most quickly at this temperature.

   At low temperature the rate is very low, it gets higher as the temperature rises

   then falls again above 50 °C.

   .... (3 marks)

① The student only got one mark for his answer to **1.2**. What else should the student have included to complete the explanation? 🖉

Remember that enzyme molecules and fat molecules must collide for a reaction to occur.

....

....

....

....

② Use the graph to check the student's answer to **1.3**. What is the correct answer? 🖉 ....

③ The student got no marks for his answer to **1.4**. Write 🖉 your own answer to **1.4** on a separate piece of paper.

Look at the command word.

What happens as the temperature increases towards the optimum?
What happens when the temperature increases further?

Your turn!

It is now time to use what you have learned to answer this exam-style question.
Remember to read the question thoroughly, looking for clues.
Make good use of your knowledge from other areas of biology.

Exam-style question

1   A student investigated the effect of pH on the activity of an enzyme that digests starch.
The student recorded how long it took for all the starch to be digested.
The results are shown in the table.

pH	2	5	7	9
Time for complete digestion in seconds		305.0	180.0	280.0
Rate of reaction (s^{-1})	0	0.33	0.55	0.36

1.1   Describe how changing the pH affects the rate of this reaction.   You are asked to 'describe' not to 'explain'.

....

....   **Remember** In the exam you can write on the question paper. On the table write 'high' where the reaction is fast and 'low' where the reaction is slow.

....

....

.... **(3 marks)**

1.2   Which pH gave the fastest reaction?

.... **(1 mark)**

1.3   Explain why the reaction was fastest at this pH.

....   Use the word 'because' in your answer. Most questions about enzymes need the words 'active site' in the answer.

....

.... **(2 marks)**

1.4   No time was recorded for pH 2. This is because the starch was not digested.
Explain why the enzyme was unable to digest the starch at pH 2.   **Remember** When the pH is a long way from the optimum, the enzyme's active site is altered.

....

....

.... **(2 marks)**

Use this checklist to check ⊘ your answers.

Checklist In my answer do I ...	⊘
include key terms such as 'active site'?	
refer to values from a graph or table if asked to interpret data?	
explain rates of reaction in terms of numbers of molecules and collisions?	
explain the effect of pH and temperature in terms of enzyme shape?	

Biology

Need more practice?

In the exam, questions about enzymes could occur as:
- simple standalone questions
- part of a question on how cells, tissues and organs work in plants and animals
- part of a question about a practical test.

Have a go at this exam-style question.

Exam-style question

1 Enzymes are often called biological catalysts.

   1.1 What is a catalyst? ....

   .... **(1 mark)**

Samples of the enzyme amylase were heated to different temperatures and mixed with starch. The rates of reaction were measured and then plotted on the graph.

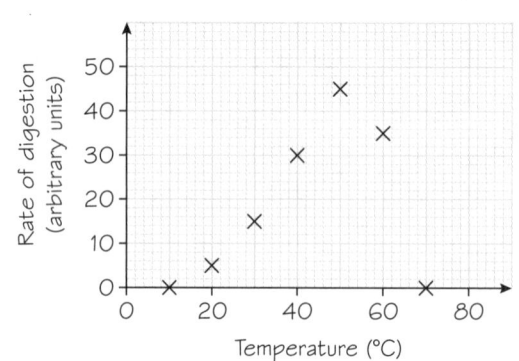

   1.2 Draw a curved line through all the points on the graph. **(1 mark)**

   1.3 Estimate the rate of starch digestion at 35 °C.

   .... **(2 marks)**

   1.4 The rate of digestion at 70 °C is lower than the rate at 50 °C. Explain why.

   ....

   ....

   .... **(2 marks)**

Boost your grade

You need to know about the enzymes carbohydrase (also known as amylase), protease and lipase and their substrates. You will also need to be able to calculate rates of reaction from data in a graph.

How confident do you feel about each of these **skills?** Colour in the bars.

1 How can I describe the action of enzymes?

2 How can I explain the action of enzymes?

3 How can I explain the effects of the conditions on enzyme action?

③ Cell division

This unit will help you to recognise when cells divide by mitosis and when they divide by meiosis. It will also help you to understand the importance of cell division in the cell cycle.

In the exam you will be asked to tackle questions such as the one below.

Exam-style question

1   Mitosis and meiosis are types of cell division.

1.1   Complete the table to show which of the features are produced by mitosis and which are produced by meiosis.

Feature	Mitosis or meiosis?
Production of egg cells	
A lizard growing a new tail	
Production of pollen in a flower	
Cells replaced on the skin to heal a cut	

(4 marks)

1.2   Identify the organs that produce gametes (sex cells) in a man and in a woman.

A man ....

A woman ....   (2 marks)

1.3   Describe two differences between mitosis and meiosis.

....   (2 marks)

You will already have done some work on mitosis and meiosis. Before starting the **skills boosts**, rate your confidence in each area. Colour in the bars.

① How can I identify the stages in the cell cycle?

② How can I describe situations where mitosis is occurring?

③ How can I explain the importance of meiosis?

The topic of cell division covers the cell cycle, mitosis and meiosis.

Cells divide in a series of stages called the **cell cycle**. First, a cell grows larger and makes more sub-cellular structures like mitochondria (for energy production) and ribosomes (for making proteins). The cell then makes copies of its chromosomes. One copy moves to each end of the cell, and the nucleus divides. The cell then divides into two new cells. Two cells are produced from one during **mitosis**.

(1) Number ✎ these statements (1-3) in the order in which they occur in the cell cycle.

Stage	Correct order
The cell increases in size and increases the number of sub-cellular structures such as ribosomes and mitochondria. DNA replicates to form two copies of each chromosome.	
The cytoplasm and cell membrane divide to form two identical, daughter cells.	
A set of chromosomes moves to each end of the cell and the nucleus divides.	

Human body cells have 46 chromosomes. The nucleus of a cell copies the chromosomes to **double** that number during **interphase**. The cell then moves a copy to each end of the cell during **mitosis**. The cell then divides during **cytokinesis**, making two new cells each with 46 chromosomes.

(2) Complete ✎ the numbers inside the cells to show what happens to the number of chromosomes during mitosis.

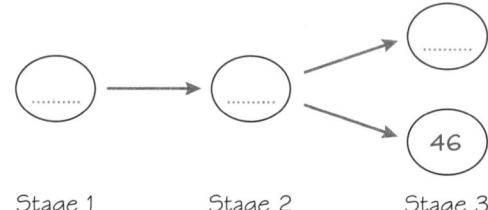

Stage 1   Stage 2   Stage 3

(3) **a** Describe ✎ what is happening inside the cell between stages 1 and 2.

....

....

b Describe ✎ what is happening inside the cell between stages 2 and 3.

....

....

Meiosis is a different type of cell division where four cells are made from one cell. In females, meiosis takes place in the ovaries, where it produces eggs. In males, meiosis takes place in the testes, where it produces sperm. The four cells that are made from the original parent cell each have a different half set of chromosomes.

(4) Complete ✎ the numbers inside the cells to show what happens to the number of chromosomes during meiosis.

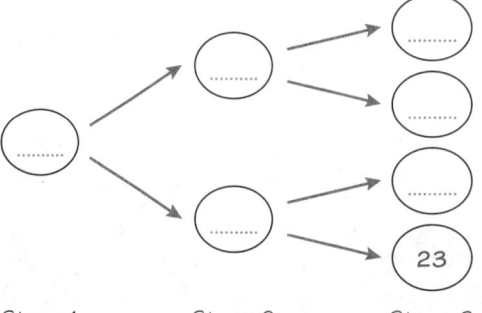

Stage 1   Stage 2   Stage 3

First meiosis makes two cells, each with a full set of chromosomes.

These divide again to make four cells, each with a half set of chromosomes.

1 — How can I identify the stages in the cell cycle?

Actively dividing cells go through a series of stages called the cell cycle. The first stage is **interphase**, where the cell makes new components and a copy of each chromosome. Then, during **mitosis**, the chromosomes move apart and the nucleus divides. The cell cycle ends with **cytokinesis** (cell division). The two new cells that are produced are genetically identical and are called **daughter cells**.

1 a Match the overall stages of the cell cycle with the more detailed descriptions.

Interphase

Mitosis

Cytokinesis

One set of chromosomes is pulled to each end of the cell, and the nucleus divides.

The cell membrane and cytoplasm divide to form two new cells.

The cell increases in size and produces more ribosomes and mitochondria. The cell also makes a complete copy of the DNA.

b The cell cycle in a tomato plant tip cell lasts 6 hours.

i Convert 6 hours to minutes.

....

ii Work out the number of minutes represented by a 1° angle on the pie chart.

....

Interphase

320°   20° Mitosis

20° Cytokinesis

iii Use the information in the pie chart to complete the table and calculate the time taken for each stage of the cell cycle.

	Angle (°)	Time in minutes	Time in hours and minutes
Interphase			
Mitosis			
Cytokinesis			

How many cells will there be after 6 h?

2 Tomato plants have 10 chromosomes in a normal cell.

a How many chromosomes are there in a cell 5 h 20 min after the start of the cell cycle?

Look at where the cell cycle will be after 5 h 20 min.

....

b How many chromosomes are there in a cell 6 h after the start of the cell cycle?

....

3 Look at this graph. It shows the stages of the cell cycle for a different organism over a 10 hour cycle.

a Which letter represents the genetic material doubling during interphase? ....

b Which letter represents the genetic material moving apart and the nucleus dividing during mitosis?

....

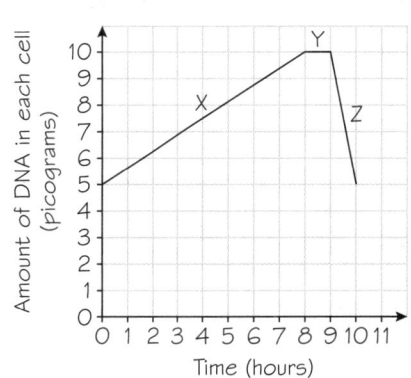

Amount of DNA in each cell (picograms) vs Time (hours)

Biology

How can I describe situations where mitosis is occurring?

Mitosis is used for increasing the number of cells during growth, when replacing damaged cells and for asexual reproduction. Mitosis produces genetically identical cells. This means that all cells in the body have exactly the same set of chromosomes.

Asexual reproduction leads to offspring produced from only one parent. All the offspring are identical and are known as clones.

Genes are sections of DNA found on chromosomes. Humans have 46 chromosomes in the nucleus of normal body cells. The nucleus controls the chemical reactions inside the cell.

(1) Look at the diagram and number 🖉 the parts of the cell in order of size with 1 as the smallest and 3 as the largest.

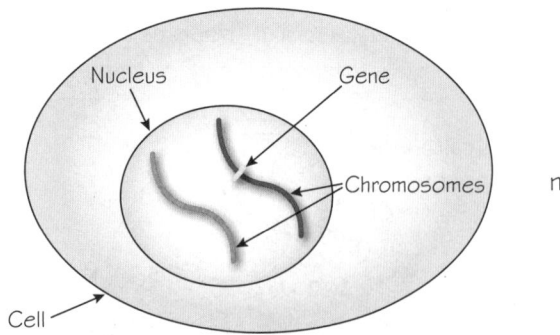

nucleus ☐   gene ☐   chromosome ☐

(2) Human skin cells A, B, C and D in the diagram have just been produced to replace some damaged cells.

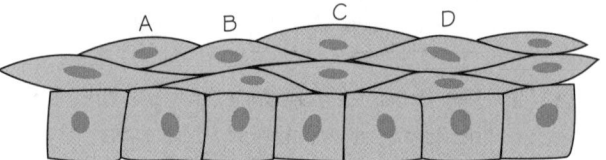

a Name 🖉 the type of cell division that has produced these new cells.

....

b What happens to the genetic material before the cell divides? 🖉

....

c How many chromosomes will be in cell A? 🖉

....

d Why is it important that skin cells can divide? 🖉

....

(3) The diagram shows a nucleus just before the nucleus begins to divide during mitosis. Complete 🖉 the second diagram to show what the nucleus of a cell produced by this mitosis looks like.

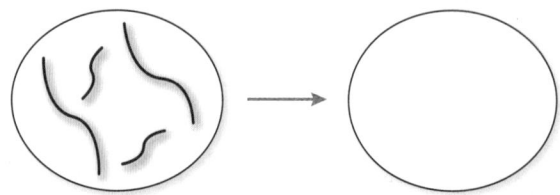

You could use this to help you remember what mitosis does.

Makes	Makes
I	Identical
Toes	T
O	Offspring
Skin	S
I	I
S	S

Remember Mitosis produces identical cells. This means that the cells will have an identical number of chromosomes.

3 **How can I explain the importance of meiosis?**

Meiosis is the type of cell division that makes sperm cells and eggs. Meiosis involves two divisions. First, two cells are made with full sets of chromosomes. These two cells then divide to make four **non-identical** cells which can be used in sexual reproduction. Each **gamete** contains half of the chromosomes needed to make a full set. They join together during fertilisation to form a **zygote.**

Gamete A sex cell such as egg or sperm. Gametes are formed by meiosis.

1 Which diagram represents cell division by meiosis? Circle Ⓐ one letter.

You could use this to help you remember what meiosis does.

Makes
Eggs
I
O
Sperm
I
S

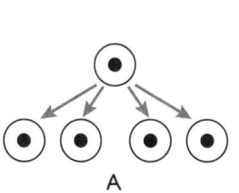

A

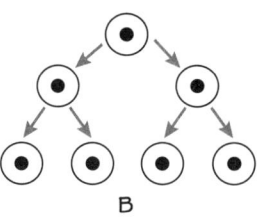

B

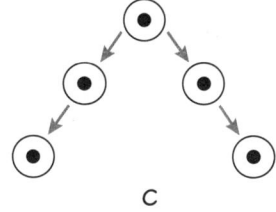

C

2 Name 🖉 an organ where meiosis takes place.

Alleles are different versions of the same gene. For example, one chromosome could carry the allele for blue eyes and the other chromosome could carry the allele for brown eyes.

3 In this diagram the top cell contains two alleles for two different genes. B is the gene for brown eyes and b the gene for blue eyes. F is the gene for brown hair and f is the gene for blond hair.

Write 🖉 the letters in the boxes to make four different combinations of alleles from these two genes, just like during meiosis.

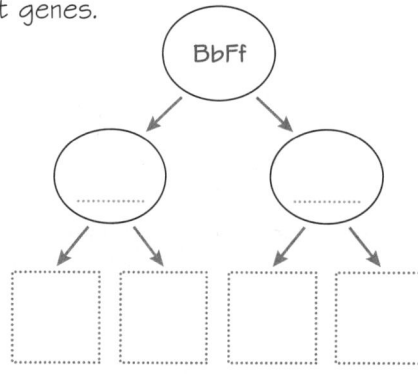

The **diploid** number in human cells is **46** chromosomes. The **haploid** number is **23**. Most body cells contain 46 chromosomes, but eggs or sperm cells only contain 23 chromosomes.

Diploid comes from the Greek for 'double' and **haploid** means 'half'.

4 Circle Ⓐ the correct keywords in this passage.

Meiosis **doubles / halves / triples** the number of chromosomes and leads to **identical / non-identical / cloned** cells.

In meiosis the cell divides twice. The first division produces two cells with the same number of chromosomes as in the original full set in the parent cell (called the **triploid / diploid / haploid** number). The second division divides those two cells and reduces the number of chromosomes to half the number in the original parent cell. The four cells now have the **triploid / diploid / haploid** number of chromosomes. This reduction is essential for **sexual / asexual** reproduction and **increases / maintains / decreases** genetic variety.

Biology

Sample response

Your understanding of mitosis, meiosis and the cell cycle will often be tested in the context of living things. Read this question carefully, use your knowledge and consider your response.

Look at this exam-style question and the answers given by a student.

1.1 Mitosis and meiosis are types of cell division. Complete the table below to show whether mitosis or meiosis is being described. Place **one** tick in each row.

Example of cell division	Mitosis	Meiosis
Gametes made in the testes	✓	✓
Growth of an embryo		
A yeast cell budding to produce offspring	✓	
Non identical cells are produced		✓

(4 marks)

1.2 The diagram shows a cell dividing.

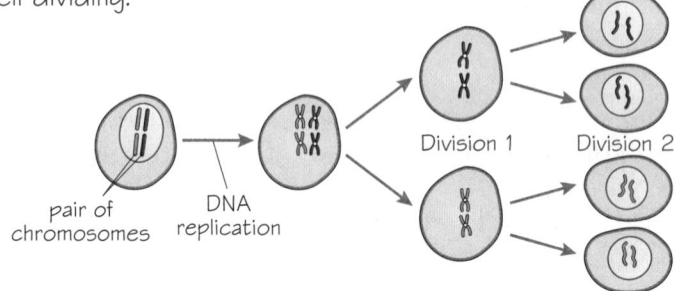

pair of chromosomes   DNA replication   Division 1   Division 2

Identify the type of cell division shown. Give three reasons for your decision.
Meiosis because the cell divides twice to produce four cells.
....

(3 marks)

1.3 Growth involves mitosis. Explain why growth does not involve meiosis.
Meiosis produces cells that are not identical. The cells produced do not have

the same genes and therefore they are not the same. The original cell has

46 chromosomes but the cells produced by meiosis have only 23 chromosomes,

and the chromosomes are all different.

(3 marks)

1 Give ✏ **two** reasons why this student did not get all four marks for **1.1**.

....

....

2 How could the student have achieved more marks for **1.2**? ✏

....

....

3 The student scored 2 marks for **1.3**. What extra response could have achieved the third mark for **1.3**? ✏

....

Your turn!

Now use what you have learned to answer this question.
Remember to read the question thoroughly, looking for clues.
Make good use of your knowledge. Read each feature carefully, use the additional guidance below and apply your knowledge from other areas of biology.

Exam-style question

1   Mitosis and meiosis are types of cell division.

   1.1   Complete the table to show which of the features are produced by mitosis and which are produced by meiosis.

Feature	Mitosis or meiosis?
Production of egg cells	....
A lizard growing a new tail	....
Production of pollen in a flower	....
Cells replaced on the skin to heal a cut	....

   (4 marks)

   • Eggs need to contain only half the genetic information; this is the haploid number.
   • Some animals can grow new body parts identical to the original one.
   • Pollen in plants is similar to sperm in animals.
   • Lots of new cells are made on both sides of a cut until they meet in the middle.

   1.2   Identify the organs that produce gametes (sex cells) in a man and in a woman.

   Think about which organs make sperm and eggs.

   A man ....

   A woman ....   (2 marks)

   1.3   Describe two differences between mitosis and meiosis.

   What type of cells are made? Where are they made? How many are made? What are the cells used for?

   ....

   ....

   ....

   ....

   ....   (2 marks)

Need more practice?

You need to be able to recognise where mitosis and meiosis are occurring in a given situation. Often, you will be tested on your understanding of both types of cell division in the same question.

Have a go at this exam-style question.

1 The diagram shows two types of cell division.

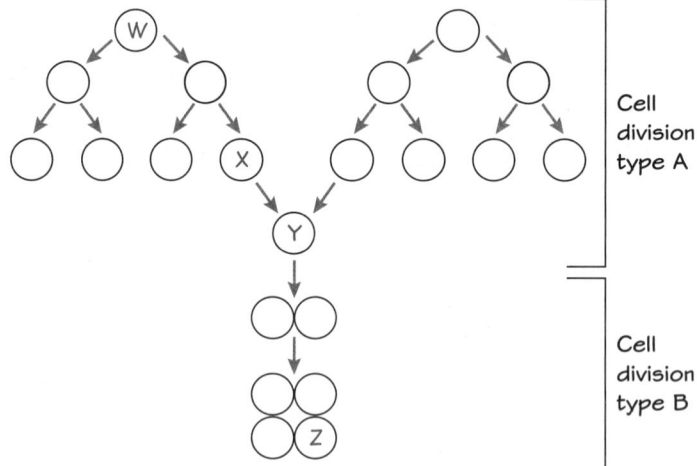

1.1 Give the name for the type of cell division labelled type A in the diagram.

.... (1 mark)

1.2 Give the name for the type of cell division labelled type B in the diagram.

.... (1 mark)

1.3 What is the name given to cell Y?

.... (1 mark)

Cell W contains 8 picograms of DNA. (1 picogram = 10^{-12} grams)

1.4 How many picograms of DNA are there in cell X?

.... (1 mark)

Boost your grade

To improve your grade, make sure you can:
• understand the overall stages of the cell cycle
• recognise and describe mitosis occurring in different situations
• explain that meiosis halves the number of chromosomes but fertilisation restores a full set.

How confident do you feel about each of these **skills?** Colour in the bars.

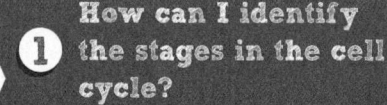

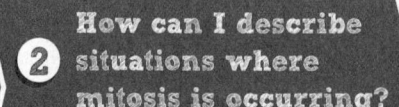

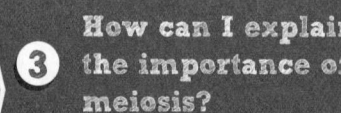

④ Practical skills

This unit will help you to answer questions based on practical work and practical situations.

Exam-style question

1   A student investigated how reaction time is affected by the time of day. This is the method:

1   The subject should be seated.

2   A ruler is held between the thumb and forefinger.

3   When the ruler is dropped the subject must catch it as quickly as possible.

4   The reaction time is measured by how far the ruler dropped before being caught.

5   This test was repeated at different times of day.

1.1   Name the dependent variable in this investigation.

....   **(1 mark)**

1.2   Give **two** variables that need to be kept constant during the investigation.

....   **(2 marks)**

The student's results are shown in **Table 1**.

Table 1

Name	Time of day	Distance ruler fell (cm)
Peter	8.30 am	12.6
Georgina	8.30 am	14.2
Anya	3.30 pm	11.5
David	3.30 pm	16.1

1.3   Explain why the results in **Table 1** do not allow a valid comparison.

....   **(2 marks)**

You will already have done some work on practical skills. Before starting the **skills boosts**, rate your confidence in each area. Colour in the bars.

1 How do I identify the independent, dependent and control variables?

2 How do I ensure my method has sufficient detail?

3 How do I draw a results table?

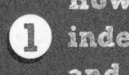

Many factors, known as variables, may affect the results of an experiment. When writing a method you must describe clearly how all these variables will be altered, measured or kept constant.

A **variable** is any factor that can change or be changed in the experiment.

- The **independent variable** is the factor you change.

- The **dependent variable** is what you measure for your results.

- A **control variable** is any factor that could affect the results of the experiment and which must be kept constant.

> The dependent variable is affected by changes in the independent variable.

Read the following extract from an experimental procedure.

1   Pour 2 cm³ of 1% catalase enzyme solution into a test tube.

2   Pour 5 cm³ of 2% hydrogen peroxide into another test tube.

3   Place both tubes in a water bath at 25 °C.

4   After 2 minutes, pour the catalase into the hydrogen peroxide.

5   Measure the height of the bubbles given off in 10 seconds.

6   Repeat at 35 °C, 45 °C, 55 °C and 75 °C.

(1) **a**   Underline (A) the independent variable and highlight (✎) the dependent variable in the box below.

> | catalase concentration   hydrogen peroxide concentration |
> | temperature   time   height of bubbles |

> The independent variable is changed and the dependent variable is measured.

b   How is the independent variable altered? (✎)

....

c   How is the volume of enzyme solution kept constant? (✎)

....

d   How is the concentration of hydrogen peroxide kept constant? (✎)

....

All your results should be put together in one results table.

- The independent variable should be in the left-hand column.

- The dependent variable should be in the right-hand column.

- Your column headings should also have the units.

(2)   Here is a results table for this experiment. Add (✎) the column headings.

> The volume of enzyme solution and the concentration of hydrogen peroxide are **control variables**. Can you think of any other variables that should be kept constant in this experiment?

25	10
35	20
45	55
55	50
75	0

1 How do I identify the independent, dependent and control variables?

Variables are any factors that can change or be changed in the experiment. There will be one independent variable, one dependent variable and many other variables (the control variables) that should be kept constant.

The **independent variable** is the variable or factor you change. It is unaffected by other variables.

1 Circle Ⓐ the independent variable in the following experiments.

   a An investigation into the effect of substrate concentration on the activity of amylase.

   b An investigation into the effect of pH on the breakdown of protein.

> The independent variable is the effect you are investigating. It is the variable you deliberately decide to change during the experiment.

The **dependent variable** is the factor that changes as a result of changing the independent variable. It is dependent upon the independent variable.

2 Underline Ⓐ the dependent variable in the following experiments.

   a Investigating the action of pectinase on the clarity of apple juice.

   b The effect of salt concentration on the length of a potato chip.

> The dependent variable is what you measure and gives you your results in the experiment.

A **control variable** must be kept constant. Control variables will be all the other variables except the independent and dependent variables. Control variables may include pH, temperature, light intensity, volume of enzyme, volume of substrate, carbon dioxide concentration, and so on.

> Control variables are factors that may affect the results. There will be many variables you need to keep constant. In an exam question do not repeat control variables that have already been described.

3 In each experiment, write ✎ **two** variables that should be kept constant.

   a Investigating how light intensity affects the rate of photosynthesis.

   ....

   b Investigating how pH affects the activity of the enzyme amylase.

   ....

4 A student used the apparatus shown in the diagram to investigate the rate of water loss from a leafy twig. The student altered the speed of the air movement.

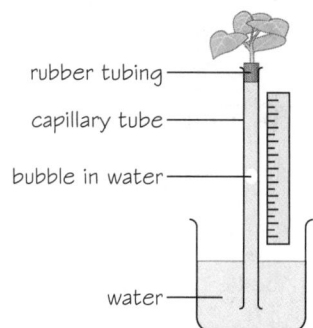

rubber tubing ———
capillary tube ———
bubble in water ———

water ———

   a What is the independent variable? ✎ ....

   b What is the dependent variable? ✎ ....

   c Write ✎ three variables that must be controlled.

   ....

   ....

   ....

Biology

How do I ensure my method has sufficient detail?

The method for an experiment must be detailed. It should describe:
- how the independent variable is altered
- how the dependent variable is measured
- how the control variables are kept constant.

This apparatus was used to investigate the effect of light intensity on the rate of photosynthesis in pondweed.

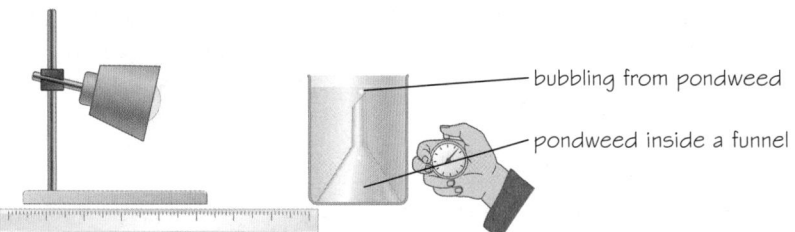

— bubbling from pondweed

— pondweed inside a funnel

The method for an experiment should be detailed enough for another student to repeat the same experiment in exactly the same way.

1 Light intensity is the independent variable. How is the light intensity altered? Tick ✓ the correct answer.

moving the light ☐   using a different light ☐

using two lights ☐   removing the light shade ☐

What is the purpose of the ruler?

2 The rate of photosynthesis is the dependent variable. How is the rate of photosynthesis measured? Circle Ⓐ the correct answer.

> counting the number of bubbles released per minute
>
> recording the volume of each bubble
>
> measuring the time between each bubble
>
> measuring the depth of water in the beaker

What is being counted?

3 A student listed the variables that must be kept constant. In each case describe 🖊 how the variable could be kept constant.

Adding a little sodium hydrogen carbonate to the water will release CO_2.

a Time for counting of bubbles ....

....

b Concentration of CO_2 ....

....

4 Complete 🖊 the sentences to describe the method.

Note that the steps in a method are often lettered or given numbers rather than written as continuous text.

1   Set up the apparatus as shown in the diagram.

2   Place a lamp at 5 cm from the .... and cover with a glass funnel.

3   Leave the pondweed for .... minutes to start photosynthesising constantly.

4   Count the number of bubbles given off each ....

5   Repeat steps .... and .... with the lamp at distances of 10, ..., ... and 40 cm.

3 How do I draw a results table?

All the results from a practical should be recorded in one results table. The format of the table is very important.
- The **independent variable** always goes in the **left-hand column** of a results table (or on the top row).
- The **dependent variable** always goes in the **right-hand column** of a results table (or on the lower row).
- The **units** should always go in the column headings, not in the main part of the table.

1. A student carried out an investigation to test the effect of temperature on the time taken to digest protein. She carried out three trials and calculated a mean.

   Complete ✏ the column headings in the table. Remember to include the units.

   What is the usual unit of temperature?

   What is the usual unit of time?

		Trial 1	Trial 2	Trial 3	mean
20					
40					

2. A student prepared the following results table for an experiment in which he tested the effect of light intensity on the rate of photosynthesis.

Rate of photosynthesis				Light intensity
Trial 1	Trial 2	Trial 3	mean	

   You need to work out which is the independent variable and which is the dependent variable.

   a. Circle Ⓐ the independent variable in the question.

   b. Give ✏ **two** things that are wrong with this table of results.

   Which variable should be in the first column?

   ....

   ....

3. A student carried out an investigation to determine the effect of changing the temperature on the rate of water loss from a leafy shoot. The loss of water was measured by weighing the plant. The student carried out three trials at 10 °C, 20 °C and 30 °C, and calculated a mean for each temperature.

   a. Circle Ⓐ the independent variable.

   b. Underline Ⓐ the dependent variable.

   c. Draw ✏ a results table for this investigation.

Biology

Sample response

Your practical skills will usually be tested in the context of a real practical that has been described. Read the question carefully, use your knowledge and consider your response.

Look at this exam-style question and the answers given by a student.

Exam-style question

1 A student investigated the effect of concentration on the rate of diffusion.

This was the method:

1 Take five Petri dishes containing colourless agar gel.

2 Use a cork borer to make a small well (hole) in the centre of the agar.

3 Pour each dilution of food dye into a well and leave for 20 minutes.

4 Measure the spread of food dye.

1.1 What is the independent variable in the student's experiment?

concentration

(1 mark)

1.2 What additional detail should the student have included in the method?

the range of concentrations used

(1 mark)

After 20 minutes, the agar dishes looked like this:

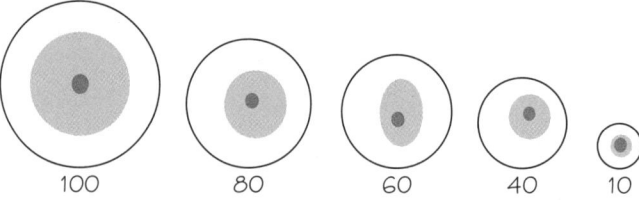

1.3 Complete the results table.

Concentration as % of original	100	80	60	40	10
Spread of dye	22	17	15	12	6

(4 marks)

1.4 Explain why the result for 60% food dye is difficult to measure.

The dye was not evenly spread.

(1 mark)

1.5 Give **two** factors that should have been kept constant in this experiment.

time, temperature, thickness of agar plate, colour of dye

(2 marks)

(1) How could the student have improved their answer to **1.1**?   Concentration of what?

(2) What important detail has the student missed in their answer to **1.3**?

(3) How many variables was the student asked to give in **1.5**?   Read the question carefully so that you don't waste time!

Your turn!

It is now time to use what you have learned to answer this exam-style question.
Remember to read the question thoroughly, looking for clues.
Make good use of your knowledge from other areas of biology.

Exam-style question

1   A student investigated how reaction time is affected by the time of day. This is the method:

1   The subject should be seated.

2   A ruler is held between the thumb and forefinger.

3   When the ruler is dropped the subject must catch it as quickly as possible.

4   The reaction time is measured by how far the ruler dropped before being caught.

5   This test was repeated at different times of day.

1.1   Name the dependent variable in this investigation.

This is the thing you measure during the practical.

....   **(1 mark)**

1.2   Give **two** variables that need to be kept constant during the investigation.

Think about things that might affect the reaction time and how far the ruler falls.

....

....   **(2 marks)**

The student's results are shown in **Table 1**.

Table 1

Name	Time of day	Distance ruler fell (cm)
Peter	8.30 am	12.6
Georgina	8.30 am	14.2
Anya	3.30 pm	11.5
David	3.30 pm	16.1

1.3   Explain why the results in **Table 1** do not allow a valid comparison.

Think about whether all the possible variables were kept the same for each test.

....

....

....

....   **(2 marks)**

Need more practice?

In the exam, questions about practical skills could ask you to:
- write a method
- suggest improvements to a method
- complete a results table.

Have a go at this exam-style question. ✎

Exam-style question

1   A student investigated the distribution of daisy plants around a large tree.

   1.1   The student listed some variables. Select the variables that might be affected
         by the tree. Place a tick against the variables that might be altered by the tree.

   shade / light intensity ☐   soil pH ☐

   time of day ☐   distance from tree trunk ☐

   soil moisture ☐   mineral concentration in soil ☐   **(3 marks)**

   The student decided to investigate whether shade created by the tree affects how many
   daisies grow. She placed a quadrat in the shade and counted how many daisy plants were in
   the quadrat. She repeated the process in the sunlight further from the tree. Her results are
   recorded in the table.

	Number of daisies in quadrat			
	Quadrat 1	Quadrat 2	Quadrat 3	mean
In shade	3	1	4	
In sun	6	4	7	

   1.2   Calculate the mean number of daisies per quadrat in the shade, and the mean
         number per quadrat in the sun. Write these values, to one decimal place,
         in the table.   **(2 marks)**

   1.3   What additional detail could the student have included in her method?

   ....

   ....

   ....

   ....   **(3 marks)**

How confident do you feel about each of these **skills?** Colour in ✎ the bars.

1 How do I identify the independent, dependent and control variables?

2 How do I ensure my method has sufficient detail?

3 How do I draw a results table?

⑤ Interpreting graphs

This unit will help you to understand how to make full use of data when the data is presented in the form of a graph.

In the exam, you will be asked to tackle questions such as the one below.

Exam-style question

1   A student measured the length of the root of a bean seedling as it germinated and grew over a number of days. The results are shown in the graph.

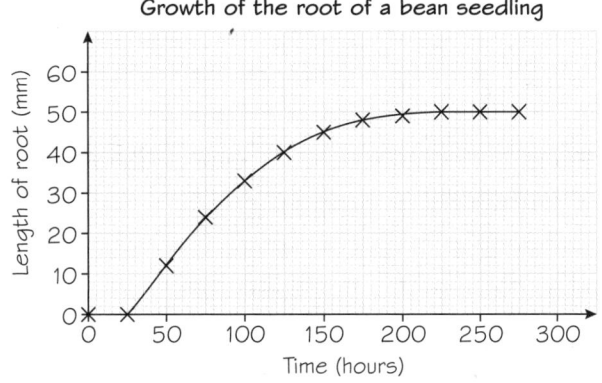

Growth of the root of a bean seedling

1.1 Describe the growth of the root over the first 100 hours.

.... (3 marks)

1.2 Give the maximum length of the root.

.... (1 mark)

1.3 Explain why the root did not appear to grow for the first 25 hours.

.... (2 marks)

You will already have done some work on graphs. Before starting the **skills boosts**, rate your confidence in each area. Colour in 🖉 the bars.

① How do I read data accurately from a graph?

② How do I describe what a graph shows?

③ How do I explain the shape of a graph?

You may be asked to describe or explain the trend shown by a graph, read data from a graph or describe the shape of a graph. You must read all the information on the graph carefully.

Use this graph for all the questions on this page.

Variation in percentage cover of reeds with distance from pond

① To describe a graph, first make sure you understand the graph.

a What does this graph show? ✎

Use the axis labels, including the units, in your answer.

....

....

....

....

....

b Describe the trend shown by the graph. Tick ✓ the sentence that gives the best description.

When asked to **describe the trend**, you need to say what happens as the value of the **independent variable** increases.

A As distance from the pond increases, the % cover of reeds increases. ☐

B As distance from the pond increases, the % cover of reeds decreases. ☐

C As the % cover of reeds increases, the distance from the pond increases. ☐

D As the % cover of reeds increases, the distance from the pond decreases. ☐

② **a** What is the percentage cover of reeds 4 metres from the pond? Circle Ⓐ the correct answer.

| 9% | 10% | 13% | 15% | 20% |

Work out what one square represents on each axis (the two axes have different scales). Then find 4 metres on the horizontal axis. Draw a vertical line up to the graph line, then draw a horizontal line across to the vertical axis.

b How far from the pond do you need to go before you stop finding reeds? Underline Ⓐ the correct distance.

| 2 m | 5 m | 6 m | 7 m | 8 m |

This question is asking you to give the distance where the percentage cover first reaches zero.

③ Explain ✎ what the data in the graph tells you about the conditions the reeds prefer.

....

....

....

....

....

....

First you need to use the data to **describe** where the reeds grow – your answer to **①** will help. Then use your biology knowledge to **explain** why the reeds grow where they do. Write a conclusion about the conditions reed plants prefer.

1 How do I read data accurately from a graph?

You may be asked to read data values from a graph. When you describe a graph, you need to include data values in your description, even if the question does not specifically ask you to.

Before you answer any questions about a graph, read the title and the axis labels, including the units. Then think about what the graph is showing you.

This graph shows the results of an investigation. The rate of reaction is measured in cubic centimetres of oxygen gas released per second (cm³/s).

Effect of substrate concentration on rate of reaction

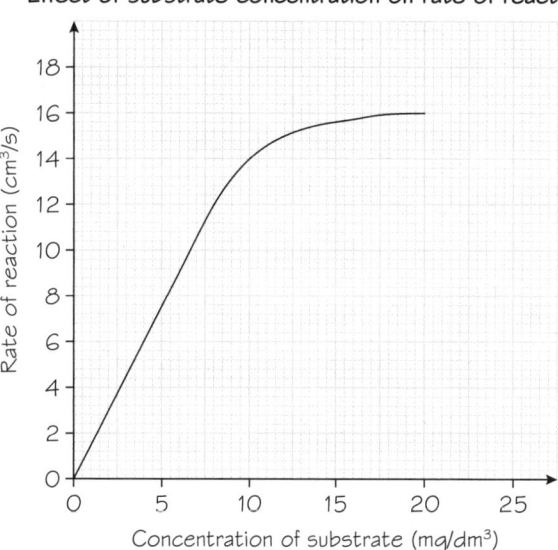

1 **a** What is the maximum rate of the reaction?

.... cm³/s

The 'maximum' is the highest value. Draw a line from the highest point of the graph to the 'rate of reaction' axis.

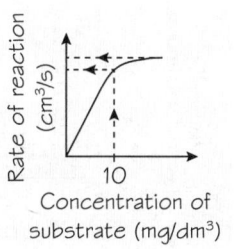

b What is the rate of reaction at a substrate concentration of 10 mg/dm³? .... cm³/s

Find 10 mg/dm³ on the 'concentration of substrate' axis. Draw a line up to the graph line, and then across to the 'rate of reaction' axis.

c What concentration of substrate produces a rate of reaction of 6 cm³/s? .... mg/dm³

2 **a** For what values of concentration of substrate is the graph a straight line? ....

b What shape is the graph for concentrations above 8 mg/dm³?

....

c Complete these sentences to describe the shape of the graph. Use data values.

Give the data values where the shape of the graph changes.

As the .... of substrate increases from 0 to 8 mg/dm³

the rate of reaction .... from 0 to .... cm³/s.

This section of the graph is a .... line, which shows the

rate of reaction is increasing at a .... rate.

For concentrations of substrate greater than 8 mg/dm³ the graph is

a .... line. The rate of reaction levels off to reach a

maximum of .... cm³/s at a substrate concentration of

.... mg/dm³.

You need to use your maths knowledge here. What does a straight line with a positive gradient tell you?

Biology

2 How do I describe what a graph shows?

A good description of a graph:
- describes how the dependent variable changes as the independent variable increases
- describes the shape of the graph and includes data values
- explains what the shape of the graph shows in context.

A graph shows how the dependent variable changes as the independent variable is changed.

(1) Look at the title and axis labels on this graph.

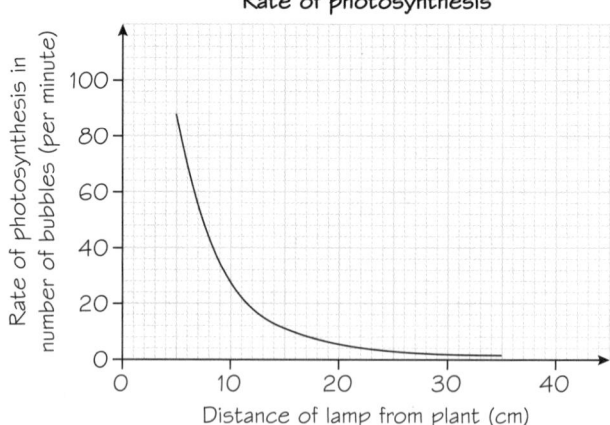

Rate of photosynthesis

(a) What is the independent variable? 🖊

The independent variable is always on the horizontal axis.

....

(b) What is the dependent variable? 🖊

Include the units with the variables.

....

(c) Circle Ⓐ the correct word in the sentence below.

As the independent variable increases, the dependent variable **increases** / **decreases**.

(d) Complete 🖊 the sentence below to describe what the graph shows.

As the distance of the lamp from the plant increases, the rate of

photosynthesis ....

'As the {independent variable} increases, the {dependent variable} …

(e) Complete 🖊 the sentence below to describe the shape of the graph. Circle Ⓐ the correct word from the choice of bold words.

The graph is a .... It **falls** / **rises** steeply for

distances from 6 cm to .... and then falls

more / **less** steeply, reaching zero at a distance of

.... cm from the lamp.

Is the graph a curve or a straight line? Where is it steep? Where does the steepness change?

(f) Complete 🖊 this sentence to explain what the shape of the graph shows. Circle Ⓐ the correct word from the choice of bold words.

The rate of photosynthesis decreases **quickly** / **slowly** for

distances from .... cm to 10 cm, and then

decreases more **quickly** / **slowly** until it reaches zero at a

distance of .... cm from the lamp.

ⓕ asks you to explain the changes in shape you described in ⓔ, in the context of the plant.

3 How do I explain the shape of a graph?

You need to use your biological knowledge to give reasons why a graph has the shape it has. There may be clues in the question to help you.

Exam-style question

1 Plants use light as a source of energy.

A student wanted to investigate the effect of light intensity on the rate of photosynthesis.

The student placed a piece of pondweed in a beaker of water and used a table lamp to provide light.

The results are shown in the graph.

1.1 Explain the shape of the curve.

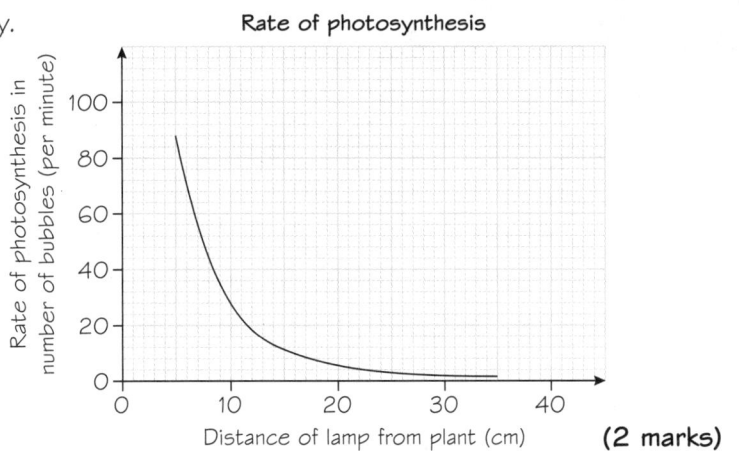

Rate of photosynthesis

(2 marks)

① What was the student investigating? ✎

....

....

② How did the student vary the light intensity? Tick ✓ the box showing the correct answer.

A By moving the lamp away from the beaker of pondweed.

B By varying the brightness of the lamp.

C By increasing the number of lamps.

D By conducting the experiment in bright sunlight and then at night.

> It is always the independent variable that is varied in an experiment.

③ Explain ✎ why plants need light for photosynthesis.

....

....

④ Complete ✎ the paragraph below to describe the shape of the curve and explain what the experiment shows.

Plants need light for .... The light is a source of ....

The higher the light .... the more energy there is for photosynthesis.

As the distance from the lamp increases, the light intensity .... and

the rate of photosynthesis .... quickly. At distances over 10 cm from

the lamp the rate of .... decreases more slowly, until it reaches

zero at a distance of .... from the lamp.

> Use your description of this graph from page 36 to help you.

Biology

Sample response

You may be expected to make use of graphs as part of a question about practical work or be asked to interpret data from a survey.

Look at this exam-style question and the answers given by a student.

Exam-style question

1   The graph shows the percentage of people dying from coronary heart disease (CHD) for smokers and non-smokers.

The probability of dying from coronary heart disease for smokers and non-smokers

Percentage of people dying from CHD (%) vs Number of cigarettes smoked (per day)

1.1 A man smokes 30 cigarettes a day. Give the probability that he will die from CHD.

20

(1 mark)

1.2 The man reduces the number of cigarettes he smokes each day from 30 to 15. How does this affect the probability that he will die from CHD?

The probability of dying from CHD decreases. It halves from 20 to 10.

(2 marks)

1.3 Describe the shape of the graph and explain what it shows.

The graph is a curve, which rises steadily. The graph shows that smoking causes cancer, as the more cigarettes smoked per day the more likely you are to get cancer. Someone who does not smoke has a 10% chance of dying but this rises to 20% if they smoke 30 cigarettes a day

(3 marks)

(1) In **1.1** the student has given the correct number. What is missing? ✎ ....

(2) In **1.2** the student is incorrect in saying the chances of dying are halved. Write ✎ the correct answer.

....

(3) How many marks would you give the student's response to **1.3**? Highlight, the parts of the answer you would give a mark. Then explain ✎ why you gave this number of marks.

....

....

....

Your turn!

It is now time to use what you have learned to answer this exam-style question.
Remember to read the question thoroughly, looking for clues.
Read the graph title and labels, and write a clear description.
Make good use of your biological knowledge.

Read the exam-style question and answer ⊘ it using the hints below.

Exam-style question

1   A student measured the length of the root of a bean seedling as it germinated and grew over a number of days. The results are shown in the graph.

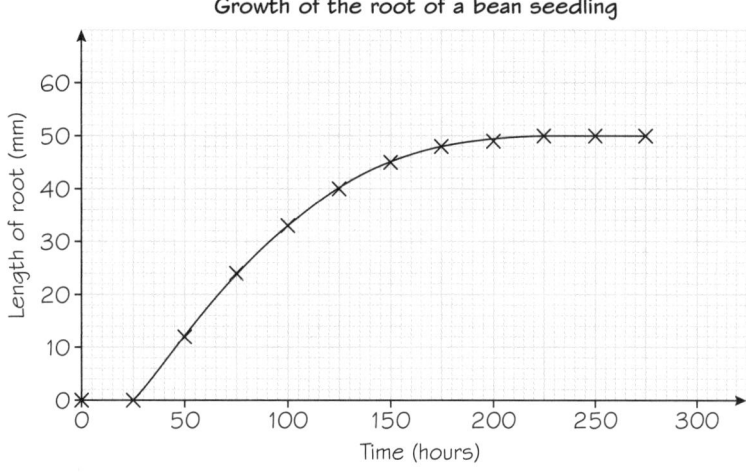

Growth of the root of a bean seedling

1.1   Describe the growth of the root over the first 100 hours.

> Describe the shape of the graph and say what this means. Give data values.

....

....

....

....   (3 marks)

1.2   Give the maximum length of the root. ....

....   (1 mark)

> Draw lines from the graph line to the axes to help you read the value accurately.

1.3   Explain why the root did not appear to grow for the first 25 hours.

....

....

....   (2 marks)

> What must happen before a seed can begin to grow? What does the seed need to absorb from the surroundings?

Biology

Need more practice?

When a question includes a graph, first read the graph title and axis labels, look at the overall shape of the graph, and work out what it is showing you. Then you are ready to answer the questions.

Exam-style question

1 The graph shows the relative risk of an accident while driving, plotted against the blood alcohol level.

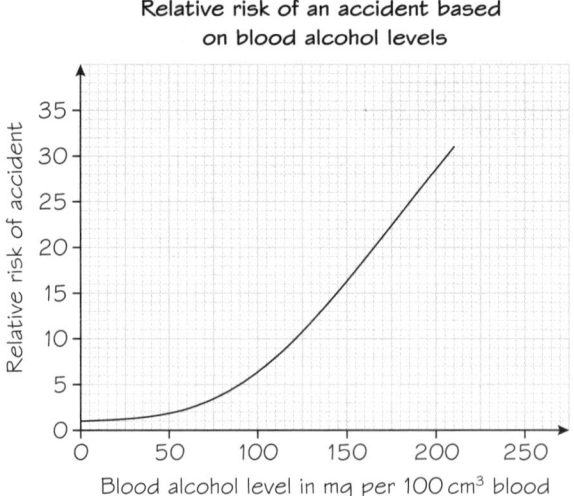

1.1 Describe how the risk of having an accident changes as the blood alcohol level increases.

Start your answer with 'As the blood alcohol level increases, ...'

Note that the command word is '**describe**'. You are not asked to *explain* why blood alcohol levels make someone a more dangerous driver.

....

....

.... (3 marks)

1.2 The legal limit for driving is a blood alcohol level of 80 mg per 100 cm³ of blood. What is the relative risk of having an accident at this blood alcohol level?

Draw lines on the graph to help you.

.... (1 mark)

1.3 The relative risk of having an accident with no alcohol in the blood is set at 1. What is the blood alcohol level when the risk is 10 times greater?

.... (1 mark)

Boost your grade

Be ready to answer questions about graphs in any question about practical work. You may also see graphs linked to questions about health. These questions may link diseases to aspects of lifestyle such as diet, obesity, smoking and alcohol consumption. In each case you may be asked to explain why the lifestyle has an effect on health.

How confident do you feel about each of these **skills**? Colour in 🖊 the bars.

1 How do I read data accurately from a graph?

2 How do I describe what a graph shows?

3 How do I explain the shape of a graph?

⑥ Maths skills

This unit will help you to understand how to carry out simple calculations, such as working out the magnification of a diagram or the rate of a reaction, and calculating percentages.

In the exam, you will be asked to tackle questions such as the one below.

Exam-style question

1   The graph shows the growth of a root plotted against time.

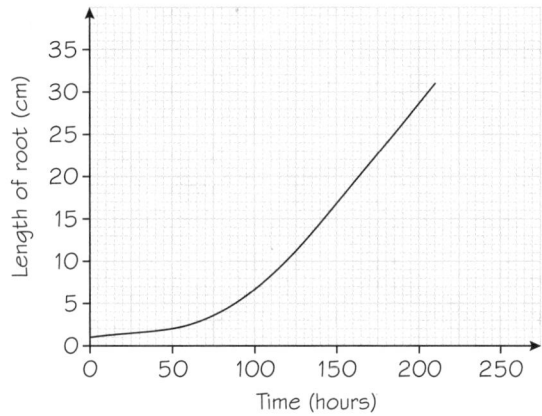

1.1   What is the length of the root at 120 hours?

.... (1 mark)

1.2   Calculate the rate of growth of the root between 120 and 150 hours.
      Show your working.

.... (2 marks)

1.3   Calculate the percentage increase in root length between 120 and 180 hours.
      Show your working.

.... (3 marks)

You will already have done some work on this topic. Before starting the **skills boosts**, rate your confidence in the maths skills needed in biology. Colour in the bars.

① How do I calculate magnification?

② How do I calculate the rate of a reaction?

③ How do I calculate a percentage?

Magnification is how many times bigger the image is than the object you are viewing.

$$\text{magnification} = \frac{\text{size of image}}{\text{size of object}}$$

① Calculate the magnification of this image by following steps **a**, **b** and **c** below.

object ☐   image

a First, write 🖉 the length of the object and the length of the image in the same units.

length of object = 6 mm

length of image = ... mm

b Now substitute 🖉 those values into the equation.

$$\text{magnification} = \frac{\text{size of image}}{\text{size of object}} = \frac{}{}$$

You might be asked to do this calculation for the **length** or the **width** of an object.

c Write 🖉 the answer.

magnification = × ....

Remember Insert a multiplication sign, ×, before the value to show it is the magnification. Magnification does not have units.

The **rate of a reaction** is the amount of product produced in one unit of time. To calculate the rate of reaction, divide the amount of product produced by the time taken.

② In a reaction, $20\,cm^3$ of oxygen is released in 5 seconds. Calculate the rate of the reaction by following steps **a**, **b** and **c** below.

a First, write 🖉 the amount produced and the time taken.

.... cm^3 produced in .... seconds.

b Divide 🖉 both sides by the same amount to work out the amount of product produced in 1 second.

$20\,cm^3$ in 5 s

÷ 5   ÷ 5

.... cm^3 in .... s

c Give 🖉 the rate of reaction.

rate of reaction = ... cm^3/s

To write one number as a percentage of another, first write the two numbers as a fraction, then convert to a decimal and then convert to a percentage.

③ In a survey of a field, 4 out of 15 quadrats contain daisy plants. Calculate the percentage of the quadrats that contain daisy plants by following steps **a**, **b** and **c**.

a Complete 🖉 the fraction: $\dfrac{4}{}$

b Convert the fraction to a decimal by dividing. Use a calculator. Complete 🖉 the fraction and decimal.

$$\frac{4}{} = \text{................}$$

c Convert 🖉 the decimal to a percentage by multiplying by 100. Round your answer to 1 d.p.

....

How do I calculate magnification?

Remember that magnification is how many times larger the image is compared to the object.

The formula for magnification is: magnification $= \dfrac{\text{size of image}}{\text{size of object}}$

Always use the same units for both measurements.

You can compare:

- the length of the object and the length of the image or
- the width of the object and the width of the image.

1 A guard cell is 0.04 mm in length. A student draws a diagram of a guard cell that is 80 mm in length. Calculate the magnification of the student's diagram by following steps **a**, **b** and **c** below.

a Write the length of the object and the image in the same units.

length of object = ... length of image = ...

b Substitute those values into the equation:

magnification $= \dfrac{\text{size of image}}{\text{size of object}} = \dfrac{\quad}{\quad} =$ ....

c Write your final answer.

magnification = × ....

Remember Always insert a multiplication sign before the value to show it is the magnification and remember that magnification does not have units.

You may be told the magnification and image size, and asked to calculate the actual size of an object. Magnification is how many times larger the image is compared to the object.

If the magnification is ×200:

- the image is 200 times larger than the object   object → ×200 → image
- the object is 200 times smaller than the image   image → ÷200 → object

2 The image of a cell is 5 mm across. The magnification is ×200. Calculate the actual size of the cell by following steps **a**, **b** and **c** below.

a Write the equation to calculate the actual size.

Use this triangle to help you rearrange the magnification equation.

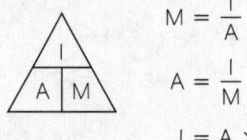

$M = \dfrac{I}{A}$

$A = \dfrac{I}{M}$

$I = A \times M$

b Substitute in the values from the question.

c Write the actual size of the cell. Remember to include the units. ....

Biology

2 How do I calculate the rate of a reaction?

A **rate** tells you how much a quantity changes in one unit of time. The unit of time can be seconds, hours, days, years, etc.

(1) Calculate the rate of photosynthesis when $24\,cm^3$ of oxygen gas is released in 3 minutes.

    a First, write ✎ the amount of oxygen produced and the time.

        .... cm^3 produced in .... minutes

    b Divide both sides by the same amount to work out the amount produced in 1 unit of time. ✎

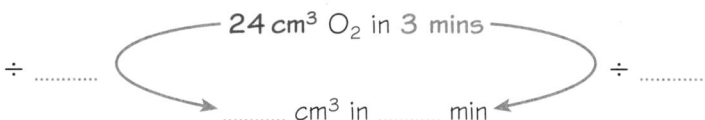

      ÷ ....   ÷ ....

                            .... cm^3 in .... min

    c Give ✎ the rate of reaction.

      rate of reaction = ... cm^3/minute

                                The quantity of oxygen is measured in cm^3 and the time is measured in minutes, so the rate is in cm^3/min.

Sometimes you may be asked to calculate the rate using information from a results table.

(2) A student investigated diffusion of food colouring in agar gel. The table shows the distance diffused from the centre of the gel over a period of time.

Time (hours)	0	1	2	3	4
Distance colouring diffused from centre (mm)	0	8	14	24	30

    a What was the rate of diffusion in the first hour? ✎ ....

    b What was the rate of diffusion in the final hour? ✎ ....

You may also be asked to calculate a rate from a graph.

(3) The graph shows the length of a root plotted against time.

    Work through the stages below to find out the rate of growth between 25 hours and 75 hours.

    a Write ✎ the increase in length and the time taken.

      .... mm growth in .... hours

    b Divide the increase in length by the number of hours. Complete ✎ the calculation.

$$\text{rate of growth} = \frac{\boxed{}\ \text{mm}}{\boxed{}\ \text{hours}}$$

    c Write ✎ the answer for the rate of growth.

      rate of growth = ... mm/hour

Graph: Length of root (mm) on y-axis (0 to 60), Time (hours) on x-axis (0 to 300).

③ How do I calculate a percentage?

'Per cent' means 'out of 100'. You may be asked to calculate a simple percentage or a percentage change.

Follow these steps to work out a percentage:

1 Write the two numbers as a fraction: $\dfrac{\text{quantity you are interested in}}{\text{total quantity}}$

2 Convert the fraction to a decimal by division, then multiply by 100.

Round the final percentage value to 1 decimal place, unless the question tells you to round differently.

① In a reaction, 80 mg of starch reduces to 24 mg. Calculate the percentage of starch that is left.

Write the fraction:
$\dfrac{\text{amount of starch at end}}{\text{amount of starch at beginning}}$

Tick ✓ the correct answer.

24.0% ☐   30.0% ☐   56.0% ☐   80.0% ☐

② In a survey of 45 692 people with type 2 diabetes, 29 435 people are overweight.

Calculate the percentage of people who are overweight.

Tick ✓ the correct answer.

35.6% ☐   45.6% ☐   64% ☐   64.4% ☐

To calculate a percentage change (increase or decrease), calculate the actual change first.

Then use this equation: $\dfrac{\text{actual change}}{\text{original value}} \times 100$

③ Calculate the percentage change when a population rises from 345 000 to 385 000.

ⓐ Calculate the actual change by subtraction. ....

ⓑ Write the original value. ....

ⓒ Substitute the two values into the equation to calculate the percentage change. Use a calculator.

A positive percentage change is a percentage increase, a negative percentage change is a percentage decrease.

④ A farmer uses a new fertiliser. His crop increases from 324 tonnes to 352 tonnes. Calculate the percentage increase in the farmer's crop.

Biology

Sample response

You may have to carry out several calculations in your exam. Calculations can be part of a theory question, or a practical question or a question on health and disease.

Look at these exam-style questions and the answers given by a student.

Exam-style questions

1 The photo shows the bacterium *Vibrio cholerae*.
   This bacterium causes cholera.

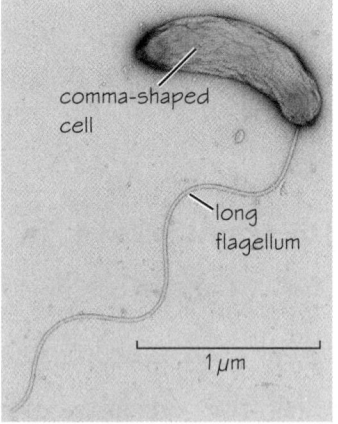

comma-shaped cell

long flagellum

1 μm

1.1 Calculate the magnification of the image.

> The image of the cell is 25 mm long. Bacteria are usually about 2 μm long.
> 25 mm = 25 000 μm. Therefore the magnification is $\frac{25\,000}{2}$ = ×12 500

(2 marks)

2 The population of the UK was 8 million in 1832 and reached 65 million by the start of 2017.

   2.1 Calculate the rate of population growth between 1832 and 2017. Show your working.

> Increase in population size = 65 million – 8 million = 57 million
> Rate of growth = $\frac{57\ million}{185}$ = 303 108.108 108 people per year

(2 marks)

① **a** In **1.1**, the student has recalled that bacteria are usually about 2 μm long and has used this information to calculate the magnification. How should she have calculated the magnification? ✎

....

b Calculate ✎ the correct magnification.   Look closely at the information provided in the photo.

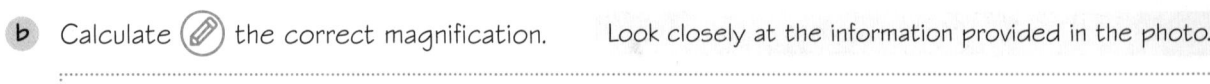

② **a** In **2.1**, the student has missed one line of working in her response. What should she add to the working? ✎

....

b Why would the student not achieve full marks for her answer? ✎   How many decimal places should there be? Can you have 0.108108 of a person?

....

Your turn!

It is now time to use what you have learned to answer this exam-style question.

Remember to read the question thoroughly, looking for clues.

Make good use of your knowledge from other areas of biology.

Exam-style question

1   The graph shows the growth of a root plotted against time.

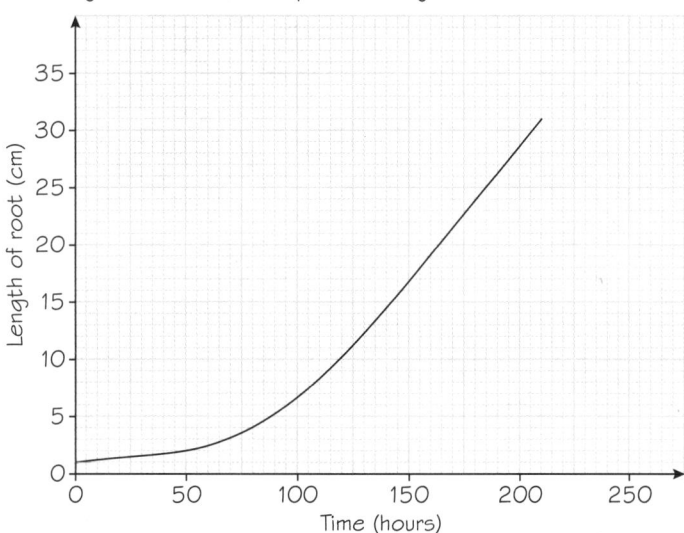

Use a ruler and draw lines on the graph to help you read values accurately.

1.1   What is the length of the root at 120 hours?

....

(1 mark)

1.2   Calculate the rate of growth of the root between 120 and 150 hours.
Show your working.

Start with .... cm in .... hours. Round your answer to a suitable number of decimal places.

....

(2 marks)

1.3   Calculate the percentage increase in root length between 120 and 180 hours.
Show your working.

- Use the formula:
  $$percentage\ change = \frac{actual\ change}{original\ value} \times 100$$
- The original value is the length at 120 hours.
- When something doubles, it increases by 100%.

....

(3 marks)

Need more practice?

You may be asked to do calculations as part of a question about practical skills, or to analyse a set of data.

Have a go at this exam-style question.

1   The photo shows the lower surface of a leaf.
    The actual size of a guard cell is 0.4 mm
    in length.

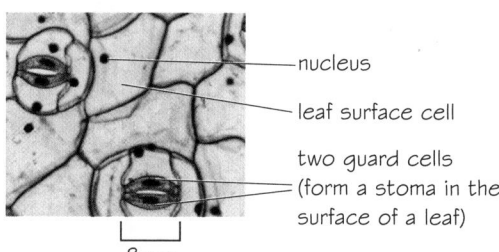

nucleus

leaf surface cell

two guard cells
(form a stoma in the
surface of a leaf)

8 mm

1.1   Calculate the magnification of the photo. Show your working.

(2 marks)

1.2   The guard cells allow oxygen produced in photosynthesis to leave the leaf.

In an investigation, 2.4 cm³ of oxygen was collected over 8 hours.
Calculate the rate of photosynthesis in cm³ oxygen per hour.

(2 marks)

1.3   When the light intensity was increased, the volume of oxygen released in 8 hours
increased to 7.2 cm³. Calculate the percentage increase in the volume of oxygen released.

(3 marks)

Boost your grade

Magnification calculations will usually be about specimens and microscope photographs.
You may be asked to do rate calculations in questions about photosynthesis practicals,
respiration, transpiration, bacterial growth, digestion or diffusion.
Percentage change calculations will often be to do with population size or health and disease.

How confident do you feel about each of these **skills?** Colour in the bars.

1   How do I calculate
    magnification?

2   How do I calculate the
    rate of reaction?

3   How do I calculate a
    percentage?

⑦ Answering extended response questions

This unit will help you to answer extended response questions by deciding what is being asked and then planning a concise answer with the right amount of detail.

In the exam, you will be asked to tackle questions such as the one below.

Biology

Exam-style question

1 Coronary heart disease can narrow the coronary arteries and reduce blood flow to the heart muscle.

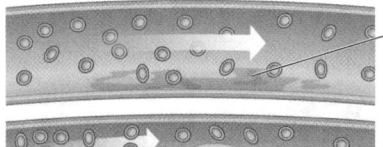 Damage to artery lining due to high blood pressure or substances in tobacco smoke

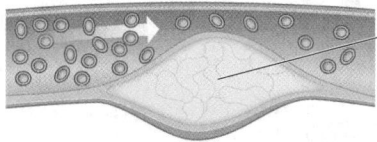 Fat builds up in the artery wall at the site of damage, making the artery narrower.

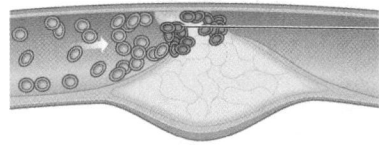 A blood clot may block the artery here, or break off and block an artery in another part of the body – causing a heart attack or stroke.

1.1 Evaluate the different ways of treating coronary heart disease.

.... (6 marks)

You will already have written some answers to extended response questions. Before starting the **skills boosts**, rate your confidence in your ability to understand, plan and write with the correct amount of detail the answer to an extended response question. Colour in 🖉 the bars.

1 How do I know what the question is asking me to do?	2 How do I plan my answer?	3 How do I choose the right detail to answer the question concisely?

The marks given for an extended response question depend both on the level of understanding shown in your answer and on how well your answer is organised.

It is essential to answer the question fully. To do this you need to look for:

- the command word

- the main topic within biology

- what aspect of the topic is being tested

- what information is supplied in diagrams and graphs.

The **command word** usually comes at the start of the question. It tells you what to do.
Here are some common command words.

- Evaluate use the information supplied, as well as your knowledge and understanding, to consider evidence for and against
- Plan write a method
- Explain give the reasons for something happening
- Describe recall some facts, an event or a process in an accurate way
- Compare describe the similarities and/or differences between things, not just write about one thing

(1) Two students (**A** and **B**) answer the following question.

Exam-style question

Explain why enzymes work more quickly when they are warm.

.... (3 marks)

A
| At high temperatures enzymes work more quickly but the molecules denature when it gets too hot. |

B
| Warmer temperatures give the molecules more kinetic energy so they move more quickly and collide more often. |

a Circle Ⓐ the command word in the exam-style question.

b Tick ✓ the answer you think is better.

c Highlight ✐ the part of the answer that you have ticked that contains the explanation.

(2) Two students (**C** and **D**) answer the following question.

Exam-style question

Compare natural selection and artificial selection.

.... (6 marks)

C
| Natural selection is selection by the environment. Individuals that are better adapted are more likely to survive and pass on their genes. The next generation are more likely to have the good adaptations. In artificial selection man selects the individuals that reproduce. It is a much quicker process as the non-adapted individuals do not breed. |

D
| Selection involves the choice of breeding partners. When farmers do this it is called artificial selection. Individuals with good adaptations breed to pass these adaptations on to their offspring so that the next generation is well adapted. This means that the population gets better and better adapted to the conditions. |

(3) a Look at the exam-style question. What does the command word 'compare' mean? ✐

....

b Highlight ✐ three words or phrases of that make answer **C** a better comparison than **D**.

1   How do I know what the question is asking me to do?

To understand what the question is asking you to do you must:
- recognise and understand the command word
- identify the topic being tested and what aspect of the topic you are being asked to write about.

(1) Circle (A) the command word in these questions:

- **a**   Describe the effect of ....
- **b**   Explain the role of predators in ....
- **c**   Draw a fully labelled diagram of a leaf.
- **d**   Calculate the rate of decomposition.

Exam-style question

1   There are many factors that contribute to the risk of developing coronary heart disease (CHD). The diagram shows how CHD can develop.

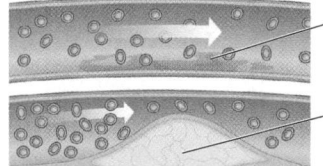
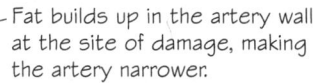
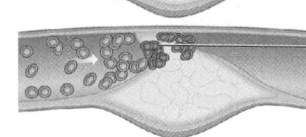

Damage to artery lining due to high blood pressure or substances in tobacco smoke

Fat builds up in the artery wall at the site of damage, making the artery narrower.

A blood clot may block the artery here, or break off and block an artery in another part of the body – causing a heart attack or stroke.

Being obese and smoking cigarettes are risk factors for developing CHD.

1.1   Explain how being obese and smoking cigarettes increases the probability that a person will develop CHD.

.... (6 marks)

(2) Read the exam-style question above carefully and analyse it by answering these questions.

- **a**   Circle (A) the **command word.**
- **b**   Circle (A) the word or words that tell you the **main topic** in biology that is being tested.
- **c**   Underline (A) the **aspects of this topic** that are being tested.
- **d**   Highlight (✐) any **useful information** in the question and on the diagram.

Exam-style question

2   Catalase is an enzyme found in many plant and animal tissues. It catalyses the decomposition of hydrogen peroxide to water and oxygen gas: $2H_2O_2 \rightarrow 2H_2O + O_2$

The oxygen gas can be collected by displacing water in an upturned test tube.

2.1   Describe how you would test a range of tissues for catalase activity and how you would ensure it is a fair test.

.... (6 marks)

(3)   **a**   Circle (A) the **command word** in the exam-style question above.

- **b**   Circle (A) the word or words that tell you the **main topic** in biology that is being tested.
- **c**   Underline (A) the **aspects of this topic** that are being tested in this question.
- **d**   Highlight (✐) any **useful information** in the question.

Biology

② How do I plan my answer?

Planning your answer is very important. First, think through the topic carefully and decide which parts of the topic are relevant to the question. Then consider the order in which points should be included.

Exam-style question

1 Homeostasis involves the maintenance of internal conditions such as temperature, pH and blood glucose concentration.

    1.1 Explain why temperature and pH need to be kept constant in the body and explain what may happen if levels change too far from the optimum.

    .... **(6 marks)**

① Circle Ⓐ the command word in the exam-style question above.   What am I being asked about?

You are told that temperature, pH and blood glucose concentration are controlled in homeostasis.

② Write ✏️ the two factors you need to focus on.   What do I know about this topic?

....

You should know a lot about how temperature and pH can affect the rate of reactions.

The first step is to recall what you can about enzyme activity. Below are the things a student jotted down about enzymes. They have underlined what they think is relevant to the exam-style question.

• Enzymes are proteins	☐	• <u>Enzymes speed up (catalyse) chemical reactions</u> 3	☐
• Enzymes have an active site	☐	• Enzymes work best at a particular pH called the optimum pH	☐
• At very high or low pH most enzymes are inactive	☐	• At low temperatures enzyme activity is low	☐
• The active site is denatured by high temperature	☐	• As temperature rises activity increases	☐
• Extremes of pH will alter the shape of the active site	☐	• This is because the enzyme molecules have more kinetic energy and collide more often	☐
• Proteases digest proteins, lipase digests fats and amylase is found in saliva and it digests starch	☐	• <u>Chemical reactions in the cell need enzymes to work quickly</u> 2	☐
• <u>Some enzymes act inside cells and others act outside cells</u> 1	☐		

③ ⓐ Highlight ✏️ the statements that explain the effect of temperature on enzyme activity.

    ⓑ Circle Ⓐ the statements that explain the effect of pH on enzyme activity.

The student has numbered the general statements about enzymes to show a possible order.

④ ⓐ Write ✏️ numbers on the plan to show what order you would write the statements about temperature.

    ⓑ Write ✏️ numbers on the plan to show what order you would write the statements about pH.

How do I choose the right detail to answer the question concisely?

You can get the right amount of detail in your answer by:
- selecting the parts of the whole topic that answer the question
- referring back to the command word to see the style you should use in your answer.

Exam-style question

1 In the early part of the twentieth century wolves were hunted and removed from the Kaibab Plateau in Arizona. The result was a rapid increase in the deer population followed by a collapse in the deer population a few years later. Scientists noted that all the trees had been stripped of leaves.

The table shows the size of the deer population on the plateau between 1900 and 1940.

Year	1900	1905	1910	1915	1920	1924	1926	1930	1940
Deer population	4000	4000	10000	24000	65000	100000	40000	20000	10000

1.1 Explain why the population of deer increased and then decreased so quickly.

.... (6 marks)

(1) What does the command word 'explain' mean? 🖉 ....

....

You must include anything you can recall that can help with your explanation. Here the population of deer rises and then falls. Think about what factors could make that happen.

Here are some student notes on the information in the table and what they already know may cause populations to rise and fall.

> A The deer population was constant before the wolves were removed. ☐
>
> B The deer population started to increase slowly, but the increase got quicker and quicker. In 1926 the population size suddenly fell. ☐
>
> C Wolves eat enough deer to limit the size of the deer population. ☐
>
> D Wolves are predators that eat deer. ☐
>
> E The population starved. ☐
>
> F The deer ate all the leaves from the trees so that there was no more food left. ☐
>
> G When the wolves were removed the deer could breed more quickly. ☐

(2) (a) Highlight 🖉 the statements that help to **explain** the rise and fall of the deer population.

(b) Number 🖉 the statements in the order you would use them in your answer.

If you can put the word 'because' in front of a statement, then it is probably an explanation.

Biology

Sample response

Extended response questions can cover any topic. Read the question carefully, use your knowledge and plan your response.

Exam-style question

1   Microbial growth can be reduced by using antibiotics or antibacterial chemicals such as bleach, by reducing the pH or by reducing the temperature.

   1.1   Evaluate each method as a way of preserving fresh meat for human consumption, making clear which method is best.

   .... (6 marks)

Here is one student's response to the exam-style question above.

> In the right conditions bacteria can grow quickly – doubling their population every 20–30 minutes. As they grow they take nutrients from their surroundings – if this is in a human they cause disease. Antibiotics such as penicillin can be used to kill the bacteria and cure the disease. However, antibiotics only kill bacteria so diseases caused by fungi and viruses are not cured. The growth of microbes can be reduced by changing the pH. When food is treated it is called pickling, but pickling would alter the taste of the meat. Another way to keep the food fresh would be salting or covering in sugar like jam. But this would give you sweet tasting meat. Meat is usually kept in the freezer or fridge. The coolness stops the microbes growing but also keeps the meat fresh without affecting the taste. This would be the best way to keep the meat as it does not alter the taste.

(1) The command word is 'evaluate'. Has the response successfully answered the question?

Circle (A) your opinion. **Yes / No**   What did the question ask?

(2) Underline (A) a sentence or phrase in the response that shows the student has evaluated the benefits and risks associated with one method of preservation.

(3) Has the student made clear which is the best method of preservation?

Circle (A) your opinion. **Yes / No**   Is the response well planned?

(4) Have all aspects of the question been considered? Circle (A) your opinion. **Yes / No**

If not, write (✎) what has been missed.

....

....

....

(5) Does the response have a logical sequence of statements?
Circle (A) your opinion. **Yes / No**   Has the student selected the
                                                           right detail for a concise answer?

Write (✎) how the sequence could be improved.

....

....

Your turn!

It is now time to use what you have learned to answer this exam-style question.
Remember to read the question thoroughly, looking for clues.
Make good use of your knowledge from other areas of biology.

Read the exam-style question and answer it using the guided steps below.

Exam-style question

1   Coronary heart disease can narrow the coronary arteries and reduce blood flow to
    the heart muscle.

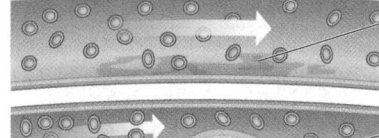
Damage to artery lining
due to high blood pressure
or substances in tobacco smoke

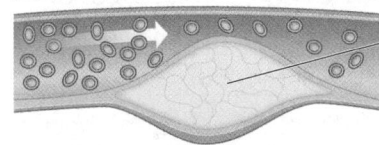
Fat builds up in the artery wall
at the site of damage, making
the artery narrower.

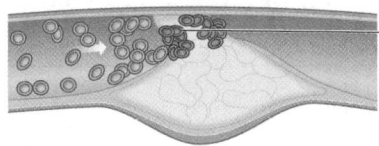
A blood clot may block the artery
here, or break off and block an
artery in another part of the body
– causing a heart attack or stroke.

1.1   Evaluate the different ways of treating coronary heart disease.

.... (6 marks)

1   Circle (A) the command word.

2   What is the disease to be treated? (✏) ....

3   On paper, copy and complete (✏) the table to answer the following questions:

   a   What different types of treatment are there?

   b   What are the benefits of each type of treatment?

   c   What are the risks associated with each type of treatment?

Treatment	Benefit	Risk

4   Now decide on the order. Should you list all the benefits of all the treatments before listing any
    risks, or would it be better to list the benefits and risks of each treatment together?
    Write down (✏) your ideas.

....

....

....

5   Write (✏) your own answer to the question on a separate piece of paper.

Need more practice?

In the exam, questions about movement into and out of cells could occur as:

- simple standalone questions
- part of a question on how cells, tissues and organs work in plants and animals
- part of a question about a practical test

An extended response question could cover any topic. Use this checklist to help you mark your answer to the exam-style question on page 55.

Checklist	
What did the question ask?	
Have you identified the main topic being tested?	
Have you written about the right aspect of the topic tested?	
Does your response answer the command word (evaluate)?	
Is the response well planned?	
Have you discussed all aspects of the question?	
Have you sequenced your statements effectively?	
Have you selected the right detail for a concise answer?	
Have you included any information that is not relevant?	
Is your answer as concise as it could be?	

You can also use this checklist to review your answer to the exam-style question below.

On paper, have a go at this exam-style question.

Exam-style question

1   You have been provided with an unlabelled sample of food.

   1.1   Describe a series of tests to find out what food groups are contained in the food.

   Ensure you provide full details of the tests you will carry out and any safety precautions.

   .... (6 marks)

Boost your grade

Ensure you know the meanings of all the command words used for exam questions.
Practise making your answers relevant by picking a question from a past paper and writing the four most important points about the topic.

How confident do you feel about each of these **skills**? Colour in the bars.

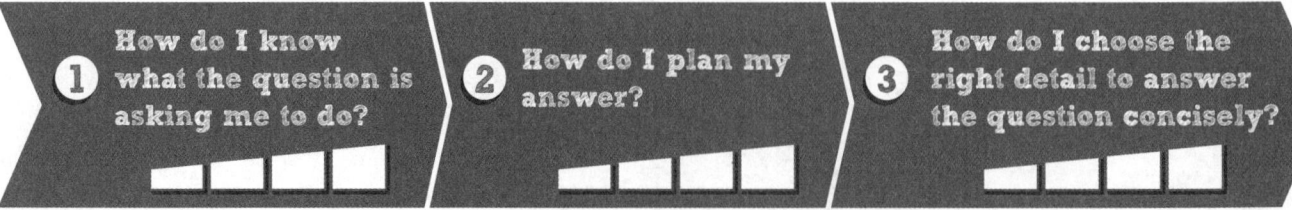

1   How do I know what the question is asking me to do?

2   How do I plan my answer?

3   How do I choose the right detail to answer the question concisely?

① Equations

This unit will help you to write and understand word and chemical equations.

In the exam, you will be asked to write, or complete, **word equations** and **chemical equations** (balanced symbol equations), and you might need to add **state symbols** to an equation. This unit will help you to write these types of equation and to get information from equations.

In the exam you will be asked to tackle questions such as the one below.

Exam-style question

1   When a small piece of sodium is added to water, a solution of sodium hydroxide is formed. Bubbles of hydrogen are seen.

1.1   Write a word equation for this reaction.

.... (2 marks)

1.2   Complete the chemical equation for this reaction.

.... + 2H$_2$O → 2NaOH + ... H$_2$O ....   (2 marks)

1.3   Give a state symbol for each of the four substances.

hydrogen   .... H ....   sodium   .... Na ....

sodium hydroxide .... NaOH ....   water   .... H$_2$O ....   (4 marks)

You will already have done some work on writing equations. Before starting the **skills boosts**, rate your confidence in writing word and chemical equations. Colour in ✏ the bars.

1 **How do I write a word equation?**

2 **How do I use state symbols?**

3 **How are symbol equations balanced?**

When a chemical reaction happens, the starting substances – **reactants** – change into new substances – **products**. An equation shows the reactants and the products in a chemical reaction.

(1) Look at the word equation for **incomplete combustion** of methane.

methane + oxygen = carbon monoxide + water

a Underline (A) all of the reactants in the equation.

b Circle (A) all of the products in the equation.

c There is one mistake in the equation. Tick (✓) the box showing the mistake.

methane reacts when burned in air, not oxygen ☐	the = should be an arrow → ☐	the incomplete combustion of methane forms carbon dioxide, not carbon monoxide ☐

(2) One **molecule** of carbon dioxide contains 1 carbon atom and 2 oxygen atoms.

Circle (A) the correct formula.

C_1O_2 CO2 CO^2 CO_2

(3) The box shows atoms and molecules of gases found in the air.

Write (✐) the formulae of the:

elements

compounds

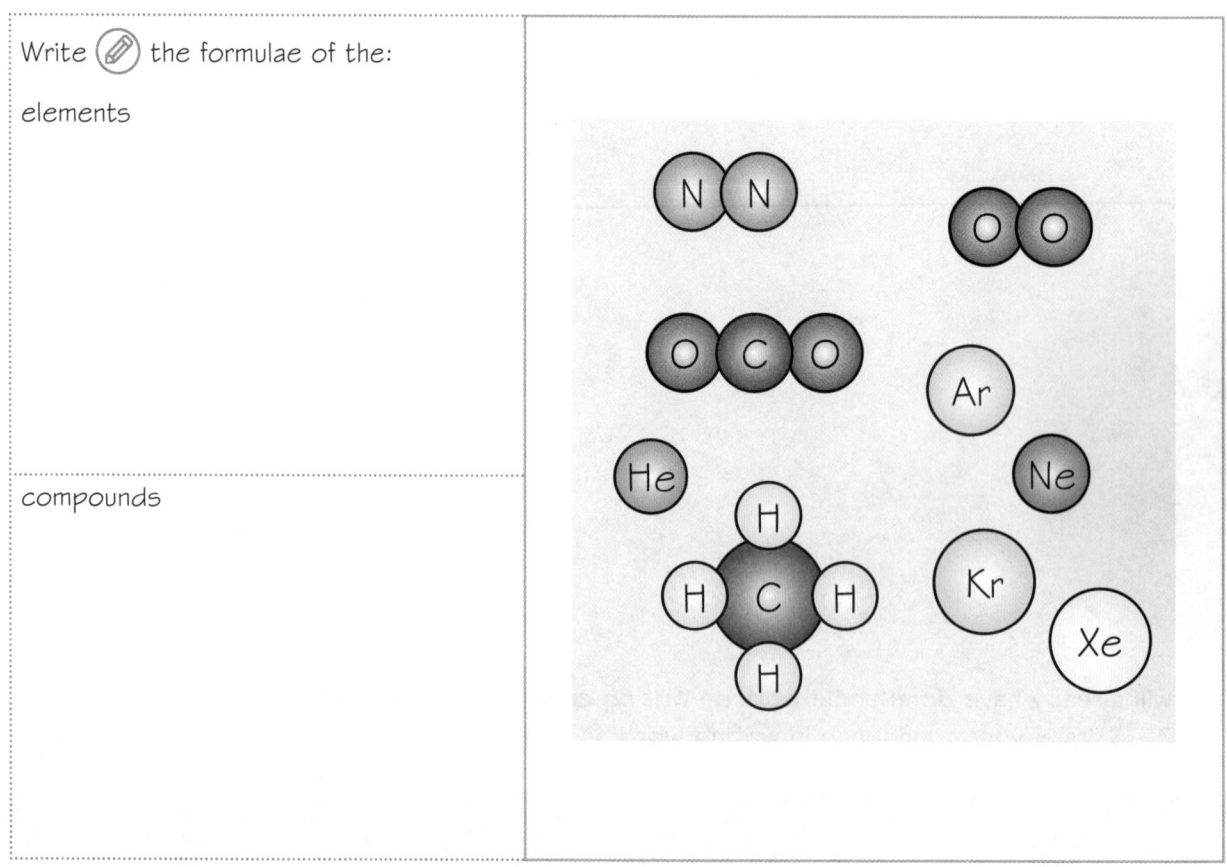

Compounds have the symbols of two or more elements in their formulae.

(4) **Aqueous** means dissolved in water. What is the state symbol for a solution in water?  ....

Skills boost

1 How do I write a word equation?

A word equation has the names of the reactants on the left, an arrow in the middle, and the names of the products on the right.

1 When sodium carbonate crystals are added to dilute hydrochloric acid, sodium chloride solution, carbon dioxide gas and water are formed.

A student writes this word equation:

| sodium carbonate crystals + dilute hydrochloric acid $\longrightarrow$ | sodium chloride solution + carbon dioxide gas + water |

Only the names of the substances should be in a word equation, not any descriptions like **solid** *or* **concentrated**.

Circle Ⓐ the **four** words that should not appear in the word equation.

2 Circle Ⓐ the correct word from each of the options below to complete the word equation for the complete combustion of octane.

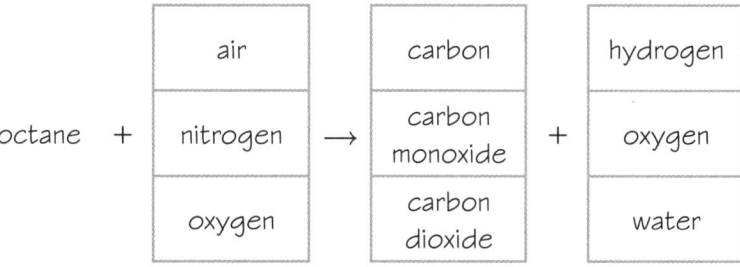

octane + [air / nitrogen / oxygen] $\longrightarrow$ [carbon / carbon monoxide / carbon dioxide] + [hydrogen / oxygen / water]

Remember When a hydrocarbon is completely combusted, the maximum number of oxygen atoms are added to each carbon atom. (Which of carbon, carbon monoxide or carbon dioxide has the most oxygen atoms for each carbon atom?)

3 The general equation for the reaction of a metal carbonate with an acid is:

metal carbonate + acid $\longrightarrow$ salt + water + carbon dioxide

Use all of the words and symbols in the box to write ✏ the word equation for the reaction of potassium carbonate with nitric acid.

| acid   carbon   carbonate   dioxide   nitrate   nitric   potassium   potassium   water   +   +   + $\longrightarrow$ |

....

....

4 The word equation for the industrial production of a metal is:

iron oxide + carbon monoxide $\longrightarrow$ iron + carbon dioxide

a Write ✏ the metal produced in this reaction. ....

The metal produced must be one of the products.

b The raw material containing this metal is called haematite.

Write ✏ the compound found in haematite. ....

The compound found in the raw material must be a reactant.

Chemistry

2 How do I use state symbols?

State symbols in an equation indicate whether the substance is a solid, a liquid, a gas or a solution in water.

(1) Join 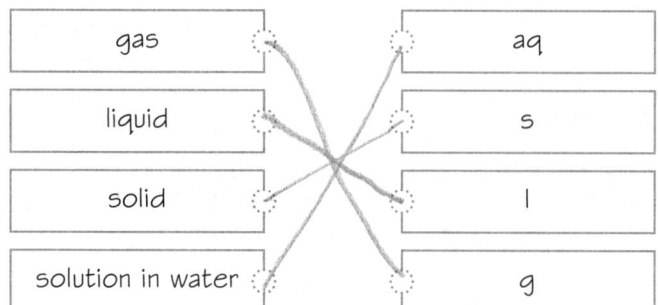 the state symbol to the description.

gas		aq
liquid		s
solid		l
solution in water		g

Aqueous means dissolved in water.

(2) When pieces of magnesium are added to dilute hydrochloric acid, a solution of magnesium chloride and bubbles of hydrogen are formed.

Key words have been highlighted to help you to choose the correct state symbol.

Add state symbols to this equation.

magnesium (Mg) + hydrochloric acid (HCl) → magnesium chloride (MgCl₂) + hydrogen (H)

(3) Copper carbonate is green. Copper sulfate is blue. Copper sulfate can be formed by adding sulfuric acid to copper carbonate. The word equation for this reaction is given.

copper carbonate (s) + sulfuric acid (aq) → copper sulfate (aq) + carbon dioxide (g) + water (l)

Complete the description of the student observations about this reaction using words from the box.

beaker	blue	bubbles	colourless	clear
disappears	evaporates	funnel	glass rod	gas
green	liquid	powder	solution	

The state symbols in the word equation will help you to choose the correct descriptions.

The acid is a .... solution and is placed in a ....

The copper carbonate, a green ...., is added and stirred with a

....

The copper carbonate .... and a ....

.... appears.

.... of carbon dioxide are seen.

3 How are symbol equations balanced?

In a chemical reaction, the atoms in the reactants are rearranged to make the products. No atoms are lost and no new atoms are formed. The numbers in the chemical equation show how many particles of each substance react.

1 Iron reacts with chlorine to make iron chloride.

   a Complete 🖉 the table.

Formula	Name of substance	Number of atoms of each element in the formula
Fe	*Iron*	Fe =
Cl_2	*Chlorine*	Cl =
$FeCl_3$	*Iron chloride*	Fe = Cl =

   b The chemical equation for the reaction is given.

$$2Fe + 3Cl_2 \rightarrow 2FeCl_3$$

Complete 🖉 the table to show the number of atoms of each element on each side of the equation.

	Reactants side of equation	Products side of equation
Fe		
Cl		

The highlighted numbers in front of each atom or molecule show how many of each there are.
So, $2FeCl_3$ = $FeCl_3$ + $FeCl_3$ = Fe + Cl + Cl+ Cl + Fe + Cl + Cl + Cl

2 Hydrogen reacts with chlorine to make hydrogen chloride, HCl. A suggested chemical equation is given.

$$H_2 + Cl_2 \rightarrow HCl$$

Which of the three students is right about balancing this equation? Tick ✓ your choice.

| this equation is already balanced ☐ | write a 2 in front of HCl $H_2 + Cl_2 \rightarrow 2HCl$ ✓ | write a 2 in the HCl $H_2 + Cl_2 \rightarrow H2Cl$ ☐ |

A chemical formula is fixed and must not be altered

3 Balance the equation by circling Ⓐ the correct number.

Mg + [1 / 2 / 3] HCl → $MgCl_2$ + H_2

Sample response

When writing equations:

- Set out in the format: reactants $\longrightarrow$ products

- In word equations, use only the chemical names of the substances

- In chemical equations, correctly write numbers in formulae, e.g. H_2O, not H^2O or H2O

- When balancing, do not change any chemical formula. Only put numbers in front of a formula.

1. Mercury and oxygen are formed when mercury oxide **decomposes**. A student has written the word equation for this reaction.

> mercury + O_2 = mercury oxide

   a There are three errors in this answer. Write down ✐ the three errors.

   i ....

   ii ....

   iii ....

   b Write ✐ the correct word equation. ....

2. Sulfuric acid has the formula H_2SO_4.

   a Identify the elements in sulfuric acid and give ✐ the number of atoms of each element in one molecule of the acid.

   Identify means 'give the name of'. You can use the periodic table to help you.

   H = ... atoms   S = ... atoms

   O = ... atoms

   b A student has written the chemical equation for the reaction between lithium and sulfuric acid to form lithium sulfate (Li_2SO_4), and hydrogen.

> $2Li + H_2SO_4 \longrightarrow Li_2SO_4 + H_2$

   Check that each formula is on the correct side of the equation, and then count the atoms on each side of the equation to see if this equation is balanced. Complete ✐ the table.

	Reactants side of equation	Products side of equation
Li		
H		
S		
O		

   c Circle Ⓐ the correct words in this sentence.

> The equation **is** / **is not** balanced because there are **the same number** / **different numbers** of atoms of each element on each side of the equation.

Your turn!

It is now time to use what you have learned to answer the exam-style question below.

Remember to read the question thoroughly, looking for clues.

Make good use of your knowledge from other areas of chemistry.

Read the exam-style question and answer it using the guided steps below.

Exam-style question

1   When a small piece of sodium is added to water, a solution of sodium hydroxide is formed. Bubbles of hydrogen are seen.

1.1   Write a word equation for this reaction.

....   **(2 marks)**

1.2   Complete the chemical equation for this reaction.

.... $+ 2H_2O \longrightarrow 2NaOH +$ ....   **(2 marks)**

1.3   Give a state symbol for each of the four substances.

hydrogen   ....   sodium   ....

sodium hydroxide   ....   water   ....   **(4 marks)**

Highlight the four substances named and put these in the correct word equation format.

1.1   ....

1.2 i   Work out and write (✐) each missing formula.

sodium   ....

hydrogen   ....

> For a metal, like sodium, use the symbol of the element (you can use the periodic table to find this). For gas elements, like hydrogen, a molecule of the gas contains two atoms (except for group 0 gases).

ii   Count the number of atoms of each element already in the equation and complete (✐) the table.

	Reactants side of equation	Products side of equation
Na		
H		
O		

iii   Complete the equation. Write (✐) a balancing number in front of the sodium atom on the left.

> If there are the same number of atoms on each side then it is balanced.

.... $+ 2H_2O \longrightarrow 2NaOH +$ ....

1.3   hydrogen   ....   sodium   ....

sodium hydroxide ....   water   ....

> Underline the words in the question that help you find the answers.

Chemistry

Need more practice?

In the exam, questions involving equations could occur as:

- simple standalone questions
- part of a question on, for example, neutralisation
- part of a question about a practical test.

Have a go at these exam-style questions.

Exam-style questions

1 Write word equations for these reactions.

1.1 Some zinc is dropped into dilute nitric acid. A solution of zinc nitrate and bubbles of hydrogen are formed.

.... (2 marks)

1.2 When silver nitrate solution is mixed with dilute hydrochloric acid, a white precipitate of silver chloride forms and a solution of nitric acid.

.... (2 marks)

1.3 Give state symbols for all of the substances in **1.1** and **1.2**.

zinc	....	nitric acid	....
zinc nitrate	....	hydrogen	....
silver nitrate	....	hydrochloric acid	....
silver chloride	....		(3 marks)

2 Complete the chemical equations by balancing them.

2.1 $Mg + \text{...........} HCl \longrightarrow MgCl_2 + H_2$ (1 mark)

2.2 $2C + O_2 \longrightarrow \text{...........} CO$ (1 mark)

2.3 $CH_4 + \text{...........} O_2 \longrightarrow CO_2 + \text{...........} H_2O$ (1 mark)

Boost your grade

To improve your grade, practise writing balanced chemical equations for the reactions you have studied in your course. Use correct symbols for the elements and chemical formulae for the compounds.

How confident do you feel about each of these **skills?** Colour in the bars.

1 How do I write a word equation?

2 How do I use state symbols?

3 How are symbol equations balanced?

② Preparing salts

This unit will help you to plan, describe and understand an experiment to prepare a salt.

In the exam you will be asked to tackle questions such as the one below.

Exam-style question

1   Copper oxide reacts with sulfuric acid. One of the products is copper sulfate.

   1.1   Describe a method of preparing a pure solution of copper sulfate.

   .... **(4 marks)**

   1.2   A student plans an experiment to prepare crystals from a pure solution of copper sulfate. This is the method used.

   Step 1: Filter the solution.

   Step 2: Pour the copper sulfate solution into a beaker.

   Step 3: Heat the solution with a Bunsen burner on a safety flame.

   Step 4: Stop heating when all the water has evaporated.

   Give an improvement for each step in this plan.

   .... **(4 marks)**

You will already have done some work on preparing salts. Before starting the **skills boosts**, rate your confidence in your ability to describe a salt preparation experiment. Colour in the bars.

1 How do I plan the method to prepare a salt?

2 How can I describe a salt preparation experiment?

3 How can I improve a salt preparation experiment?

Salts are prepared by the **neutralisation** of acids. You can tell the type of salt that will be formed from the formula of the acid used. You need to know three acids: hydrochloric, nitric and sulfuric.

1 a Match 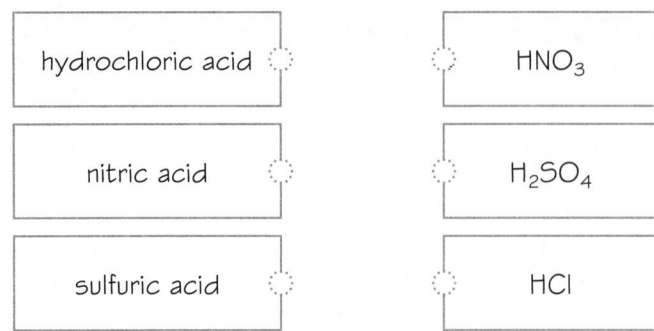 each acid with its formula.

hydrochloric acid	HNO_3
nitric acid	H_2SO_4
sulfuric acid	HCl

b Now match 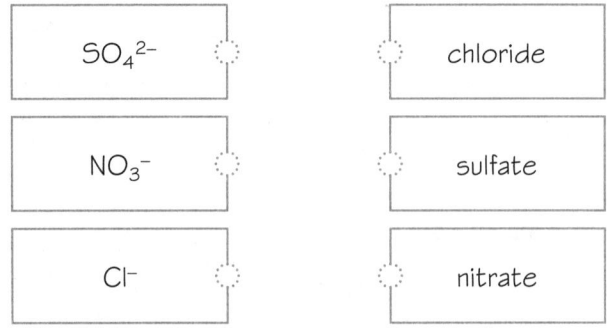 the formula with the name of an **ion** contained in the acid.

$SO_4{}^{2-}$	chloride
$NO_3{}^-$	sulfate
Cl⁻	nitrate

To know how to prepare a salt, you need to understand how an acid will be neutralised by a metal, a base or a metal carbonate.

2 Complete 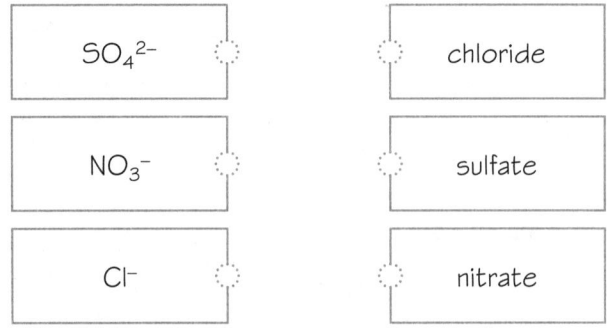 the general equations for the neutralisation of acids. Use words from the box. Check your answers. Then make sure you learn these general equations.

carbon dioxide	hydrogen	water	water

acid + metal $\longrightarrow$ salt + ...

acid + base/alkali $\longrightarrow$ salt + ...

acid + metal carbonate $\longrightarrow$ salt + ... + ...

Only the metal and the metal carbonate react with an acid to form a gas product.

3 The diagram shows apparatus that can be used when a salt is **crystallised** from a solution.

Label 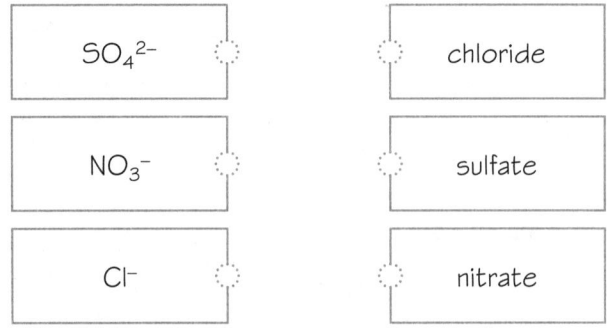 the four pieces of apparatus.

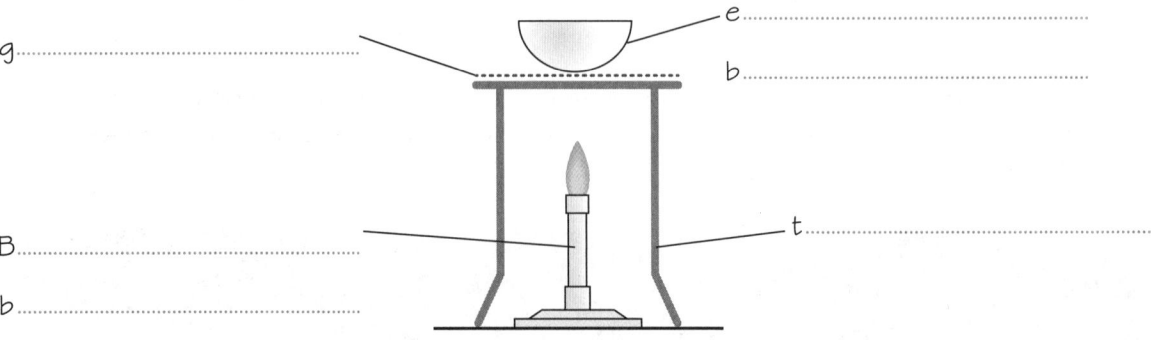

g..

e..

b..

B..

b..

t..

How do I plan the method to prepare a salt?

When preparing a salt, the acid must be neutralised. The stages are:
Stage 1 Prepare a solution of the salt by neutralising the acid.
Stage 2 Filter off the **excess** solid reactant.
Stage 3 Heat the pure solution of the salt to start forming crystals.

The general equation is: acid + metal carbonate ⟶ salt + water + carbon dioxide

1 A student has written a description of how to prepare a salt.

> Some potassium carbonate powder is added to dilute hydrochloric acid.
> Potassium chloride solution, carbon dioxide gas and water are formed.
> Pure potassium chloride crystals can be obtained from the potassium chloride solution.

In the description:

a circle Ⓐ the metal carbonate and write ✏ M.

b circle Ⓐ the salt formed and write ✏ S.

2 These are the steps in making a salt.

A	Add more potassium carbonate powder so some powder remains.
B	Place the **filtrate** into an evaporating basin.
C	Place the dilute hydrochloric acid in a beaker.
D	Filter the mixture in the beaker.
E	Warm the solution in the evaporating basin until about half the water has evaporated.
F	Add some potassium carbonate powder and stir until the powder has all disappeared.
G	Leave the evaporating basin in a warm place until crystals have formed.

Sort the steps into the correct stages and write ✏ the letters in the order that you would carry them out.

Stage 1 Prepare the solution   **Stage 2** Filter off the excess potassium carbonate   **Stage 3** Heat the pure solution of potassium chloride to start forming crystals

3 The diagram shows stage 2 of the experiment. Complete ✏ the labels.

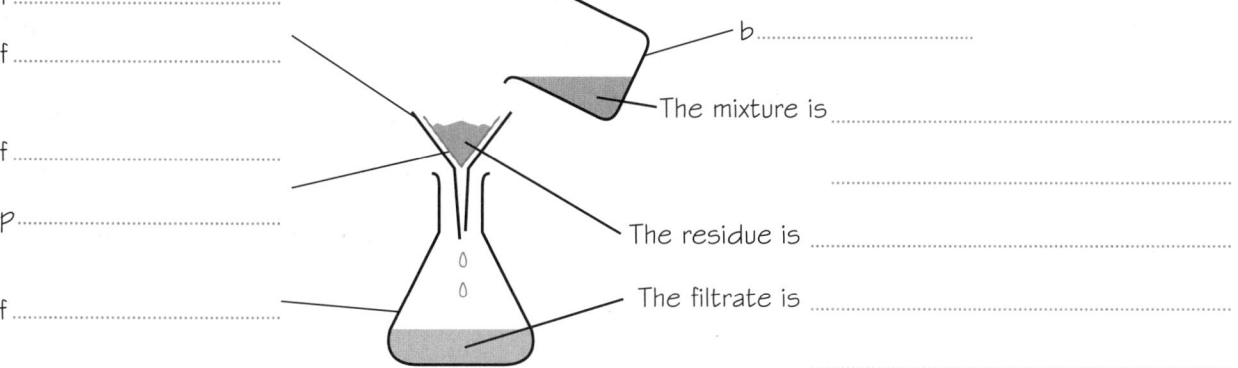

f ....

f ....

f ....

p ....

f ....

b ....

The mixture is ....

The residue is ....

The filtrate is ....

Chemistry

2 How can I describe a salt preparation experiment?

These stages in a salt preparation experiment are all needed to obtain pure, dry crystals.
Stage 1 Prepare a solution of the salt – an excess of the solid reactant is added to neutralise all the acid.
Stage 2 Filter off excess solid – this removes the excess solid reactant.
Stage 3 Crystallise the solution – the water is evaporated slowly to allow crystals to form.

① This is the method used for preparing zinc sulfate crystals.

Stage 1
Step 1: Place some sulfuric acid into a beaker and warm the acid with a Bunsen burner.
Step 2: Add some zinc oxide powder and stir until all the powder has disappeared.
Step 3: Repeat step 2 until some powder remains.

Stage 2
Step 4: Filter the mixture.

Stage 3
Step 5: Place the zinc sulfate solution in an evaporating basin and heat using a water bath until crystals have formed.

a Identify ✏ the base and the salt in this experiment.

Base: ....   Salt: ....

b Tick ✓ the reason for heating in step 1.

| a different salt is formed at room temperature | ☐ | the acid evaporates when it is heated | ☐ | the reaction is too slow at room temperature | ☐ |

c Tick ✓ the reason for adding more zinc oxide in step 3.   Zinc sulfate is soluble in water.

| so that all of the acid is neutralised | ☐ | the zinc oxide only reacts when there is an excess | ☐ | the powder that remains is the zinc sulfate salt | ☐ |

d Tick ✓ the methods that would form crystals, rather than a powder, in step 5.

| heat with a water bath until saturated | ☐ | heat with an electric heater until saturated | ☐ | put in a freezer | ☐ |
| heat with a Bunsen burner until dry | ☐ | neutralise with an alkali | ☐ | leave in a warm place | ☐ |

The solution must be warm, or be slowly heated, to form crystals.

② The table shows some ions found in bases and salts.
Circle Ⓐ the correct formula for each substance.

positive ions			negative ions		
K^+	Mg^{2+}	Na^+	Br^-	Cl^-	O^{2-}

a potassium chloride KCl K_2Cl KCl_2

b sodium oxide NaO Na_2O NaO_2

c magnesium bromide $MgBr$ Mg_2Br $MgBr_2$

Remember The charges on the positive ions must equal the charges on the negative ions. You can check the symbols used in the periodic table.

3 How can I improve a salt preparation experiment?

The method used to prepare a salt must:
- neutralise all the acid
- remove the excess solid reactant
- correctly heat the solution.

Write the improvement needed for each step in this method for making sodium sulfate crystals.

Step	Diagram	Improvements
Step 1 ① Pour 50 cm³ sulfuric acid into a beaker.	cm³ 100 80 60 40 20 cm³ 100 80 60 40 20	**a** What safety precaution should you take? .... **b** What apparatus could you use to accurately measure 50 cm³? ....
Step 2 ② Add one spatula of sodium carbonate powder and mix until all the solid disappears.		**a** How should you mix the powder and acid? .... **b** How can you tell that you have added an excess of the powder? ....
Step 3 ③ Pour the solution into a beaker.		**a** How can you remove the excess sodium carbonate? .... **b** What piece of apparatus should you put the sodium sulfate solution into? ....
Step 4 ④ Heat the solution with a Bunsen burner until all the water has boiled off.	heat	**a** What would you see if all the water is boiled off? .... **b** What could you use to heat the solution gently, so crystals form? ....

Chemistry

Sample response

When writing a plan for a salt preparation experiment:

- Measure some acid into a beaker.
- Add an excess of the solid metal oxide or metal carbonate.
- Filter off the excess solid.
- Gently heat the salt solution in an evaporating basin using a water bath or electric heater.

1 Magnesium chloride crystals can be prepared from hydrochloric acid and magnesium. Here is one student's description of the method and observations.

Pour 50 cm³ dilute hydrochloric acid into a test tube. Add a small amount of magnesium ribbon to the acid. Fizzing is seen and the magnesium disappears. Continue adding acid and stir with a spatula until some magnesium remains. There is now a residue of magnesium. Pour off the magnesium chloride solution into an evaporating basin. Heat the evaporating basin until all the water has evaporated. Leave until crystals form. Then pat the crystals dry with writing paper.

How can you accurately measure 50 cm³ acid?

How should the powder and acid be mixed?

How can the excess solid be removed?

Evaporating all the water makes powder.

a Improve the description by replacing the seven highlighted parts.

b The student's method misses out the observation of what the crystals look like.

Give a description here.

Only transition metal compounds have coloured crystals.

....

2 **a** What caused the fizzing in the reaction? A gas is formed when a metal reacts with an acid.

....

b Write a word equation for the reaction.

.... + ... ⟶ .... + ...

....

Your turn!

It is now time to use what you have learned to answer the exam-style question below.

Remember to read the question thoroughly, looking for clues.

Make good use of your knowledge from other areas of chemistry.

Read the exam-style question and answer it using the guided steps below.

Exam-style question

1   Copper oxide reacts with sulfuric acid. One of the products is copper sulfate.

1.1   Describe a method of preparing a pure solution of copper sulfate.

.... (4 marks)

1.2   A student plans an experiment to prepare crystals from a pure solution of copper sulfate. This is the method used.

Step 1: Filter the solution.
Step 2: Pour the copper sulfate solution into a beaker.
Step 3: Heat the solution with a Bunsen burner on a safety flame.
Step 4: Stop heating when all the water has evaporated.

Give an improvement for each step in this plan.

.... (4 marks)

1.1   *Stage 1: Prepare a solution of the copper sulfate.*   Describe the two stages in salt preparation using scientific terms and naming the apparatus.

....

....

....

....

Stage 2: Filter off excess solid.

....

....

....

1.2   *Step 1:* ....

What is being filtered? Is this needed?

Step 2: ....

What is the correct apparatus?

Step 3: ....

How could you gently heat the solution?

Step 4: ....

Powder is formed if the solution is heated until dry.

Chemistry

Need more practice?

In the exam, questions involving salts could occur as:

- simple standalone questions
- part of a question on neutralisation.

Have a go at this exam-style question.

Exam-style question

1 Some pure, dry calcium nitrate crystals are to be prepared.

1.1 Plan a method to prepare these crystals. A list of the available materials and apparatus is given.

• eye protection	• filter funnel and filter paper	• 250 cm³ beaker
• evaporating basin	• stirring rod	• 25 cm³ measuring cylinder
• Bunsen burner, tripod and gauze	• dilute nitric acid	• hot water bath
• spatula	• calcium carbonate	

(6 marks)

1.2 Balance the equation for the reaction.

$CaCO_3 + \ldots\ldots HNO_3 \rightarrow Ca(NO_3)_2 + H_2O + CO_2$

(1 mark)

Boost your grade

To improve your grade, make sure you know the difference between strong and weak acids, and concentrated and dilute acids. Check that you understand the pH scale and how indicators can be used. Practise writing balanced chemical equations for neutralisation reactions, and the ionic equation for neutralisation.

How confident do you feel about each of these **skills?** Colour in the bars.

1 How do I plan the method to prepare a salt?

2 How can I describe a salt preparation experiment?

3 How can I improve a salt preparation experiment?

③ Covalent molecules

This unit will help you to describe and draw the molecules of covalent substances. It will help you explain some of the properties of substances made of small molecules: in particular their low melting points and non-conductivity.

In the exam you will be asked to explain covalent bonding. You will be asked to draw diagrams of molecules, and you will need to explain why substances made of small molecules have low melting points and do not conduct electricity. This unit will help you to understand, describe and explain simple covalent substances.

In the exam you will be asked to tackle questions such as the one below.

Exam-style question

1   Ammonia, NH_3, is made of small molecules.

1.1   Describe how the atoms are held together in an ammonia molecule.

.... (2 marks)

1.2   Draw the dot-and-cross diagram of an ammonia molecule.

.... (3 marks)

1.3   The boiling point of ammonia is $-33\,°C$. The N–H bonds in ammonia are strong. Explain why ammonia has a low boiling point even though the N–H bonds are strong.

.... (3 marks)

1.4   Does liquid ammonia conduct electricity? Explain your answer.

.... (2 marks)

You will already have done some work on bonding. Before starting the **skills boosts**, rate your confidence in your ability to understand simple covalent substances. Colour in the bars.

① How can I describe bonding in substances with only non-metal atoms?

② How do I draw a dot-and-cross diagram of a molecule?

③ How can I explain the properties of a covalently bonded substance?

Chemistry

Substances can have **ionic bonding**, **covalent bonding** or **metallic bonding** holding the particles together. The type of bonding depends on whether the elements involved are metals or non-metals, as shown in the table below.

(1) The table gives some information about the four types of structure.

Structure	Types of atom	Bonds	Particles in structure	Examples
small molecules	non-metal only	covalent	molecules	chlorine, water
giant covalent lattice	non-metal only	covalent	atoms in large lattice	diamond, silicon dioxide
giant ionic lattice	metal + non-metal	ionic	ions	sodium chloride
giant metallic lattice	metal only	metallic	positive ions in a sea of electrons	copper

Write ✎ the name of the type of structure below each diagram.

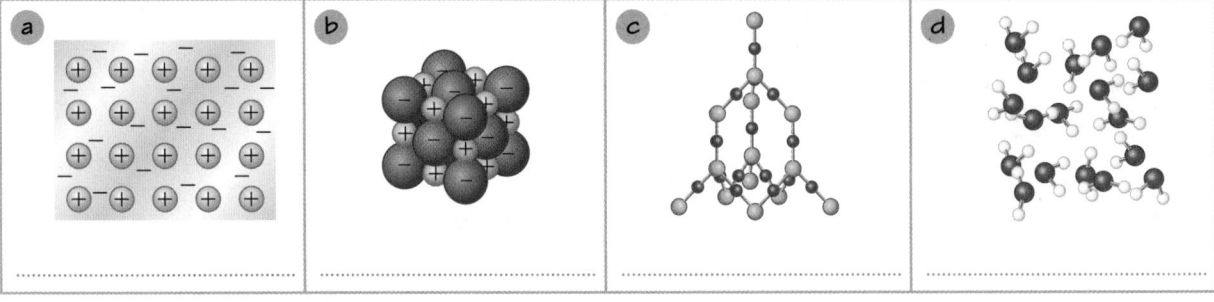

a .... b .... c .... d ....

Remember Look for the clues in the diagrams.
- The small molecules are *separate molecules*, not one giant structure like all the others.
- The giant ionic lattice has *positive and negative ions*.
- The giant covalent lattice has *many neutral atoms*.
- The giant metallic lattice has *positive ions* and a *sea of electrons*.

(2) Methane is made of small molecules.

These three diagrams give different ways of showing a molecule of methane.

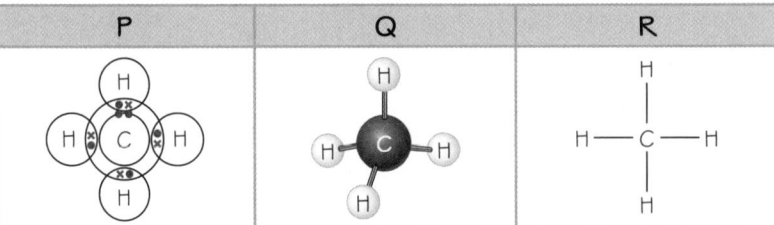

P	Q	R

a Write ✎ the formula of a molecule of methane. ....   Diagram **R** may help.

b How many electrons are there in one molecule

of methane? ✎ ....   Diagram **P** may help.

c How many covalent bonds in one molecule

of methane? ✎ ....   Diagram **R** may help.

d Which of these is an advantage of diagram **Q** compared to diagrams **P** and **R**?

Tick ✓ **one** box.

A. Diagram **Q** shows the number of bonds in a molecule.   ☐

B. Diagram **Q** shows how the electrons are arranged in a molecule.   ☐

C. Diagram **Q** shows the 3D shape of the molecule.   ☐

1 **How can I describe bonding in substances with only non-metal atoms?**

When non-metal atoms are joined in an element or in a compound, the atoms are joined by covalent bonds. The atoms **share** electrons to form a covalent bond.

(1) An atom of chlorine has 17 electrons.

a These electrons are arranged in three shells.

The first shell has a maximum of 2 electrons.

The second shell has a maximum of 8 electrons.

The third shell has a maximum of 8 electrons.

Give ✎ the number of electrons in each shell in chlorine.

First shell (inner shell)	
Second shell	
Third shell (outer shell)	
Total	17

b Write ✎ the electronic structure of chlorine: .... , ... , ...

c What is the link between the group number of chlorine in the periodic table and the number of electrons in chlorine's outer shell? ✎

Look at chlorine's group number in the periodic table.

....

....

d Complete ✎ the diagram of a chlorine atom by adding the outer shell (third shell) electrons only.

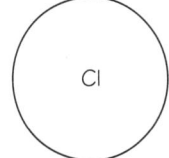

e The diagram shows the arrangement of electrons in the outer shells of a chlorine molecule.

Circle Ⓐ the covalent bond in the diagram.

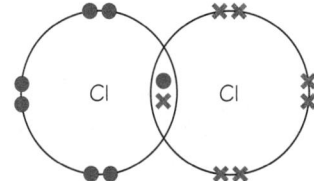

f Complete the sentences. Circle Ⓐ the correct word or number from each pair.

The chlorine atoms **share** / **transfer** a pair of electrons. The shared pair of electrons is a **covalent** / **ionic** bond. Each chlorine atom now has a full outer shell of **2** / **8** electrons.

g Look at these three molecules.

P	Q	R
O=O	H–O–H	N≡N
oxygen	water	nitrogen

Write ✎ the **name** of the substance with:

i single covalent bonds ....

ii a double covalent bond ....

iii a triple covalent bond ....

Chemistry

② How do I draw a dot-and-cross diagram of a molecule?

In a covalent molecule the atoms share electrons. Usually, after the electrons have been shared, each atom has a full outer shell. The number of electrons in a full shell is given on page 75.

① Complete the sentences. Circle (A) the correct choice.

a When most atoms form a molecule and their outer electron shell is filled, the outer shell contains **2 / 8 / 10** electrons.

b Hydrogen atoms only have one electron shell, so when they form a molecule and their outer shell is filled, it contains **2 / 8 / 10** electrons.

The number of bonds an atom will form depends on how many electrons it needs to fill its outer shell.

② The atomic number of oxygen is 8. The diagram shows the electronic structure of an oxygen atom.

Each time an atom shares, it adds one electron to the outer shell.

a Complete ⌀ the sentences.

Oxygen has .... outer electrons and is

found in group .... of the periodic table.

An oxygen atom has to share electrons .... times to get a full outer shell.

> The atomic number of an element can be found in the periodic table. The number of outer electrons is the same as the group number of the atom's element in the periodic table.

b The diagram shows a dot-and-cross diagram of a molecule of water.

Complete ⌀ the diagram by adding the hydrogen electrons. Use a cross for each electron.

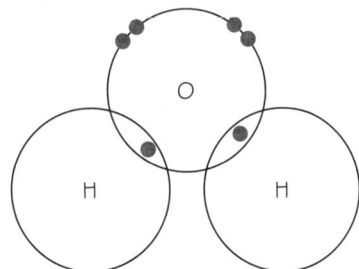

> First show the two shared electrons where the electron shells overlap.
> Now count the electrons! In a water molecule, the oxygen atom should have a full shell of eight electrons and the hydrogen atoms should each have a full shell of two electrons.

c The diagram shows the outer electron shells of a molecule of hydrogen chloride. The electron of hydrogen has been shown with a dot.

Complete ⌀ the diagram by adding the seven outer electrons of the chlorine atom. Show these electrons with a cross.

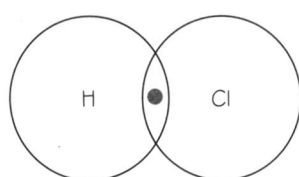

> One of the chlorine electrons must go where the electron shells overlap to show the covalent bond.

3 How can I explain the properties of a covalently bonded substance?

Covalent bonds are strong. The melting point of a covalently bonded substance depends on whether these strong bonds need to be broken when the substance melts. Covalently bonded substances do not conduct electricity as they have no free charged particles (ions or electrons).

A diamond lattice is shown on the left. Hydrogen chloride is a gas at room temperature but, if it is cooled down enough, hydrogen chloride molecules arrange themselves in a lattice as shown on the right. The forces that attract one HCl molecule to another in the solid lattice are called **intermolecular forces**.

	Diamond	Hydrogen chloride
structure	covalent bonds	covalent bond / intermolecular force
melting point	over 3500 °C	−114 °C
boiling point	over 4000 °C	−85 °C
conductor of electricity?	no	no

① What are the states of diamond and hydrogen chloride at 25 °C?

Tick ✓ **one** box for each substance.

Diamond: solid ☐ liquid ☐ gas ☐

Hydrogen chloride: solid ☐ liquid ☐ gas ☐

> If the temperature is below the melting point, a substance will not be hot enough to melt to form a liquid. Only if the temperature is above the boiling point will it be hot enough for a substance to boil to form a gas.

② Which is the correct description of the two structures at room temperature? Tick ✓ **one** box.

A. Diamond and hydrogen chloride both have giant lattices. ☐

B. Diamond and hydrogen chloride are both made of small molecules. ☐

C. Diamond has a giant lattice but hydrogen chloride is made of small molecules. ☐

If covalent bonds have to be broken when a substance is melted, lots of energy is needed to break the bonds because covalent bonds are strong. If a substance is made of small molecules, only the intermolecular forces have to be broken, which are much weaker than covalent bonds.

③ Which is the correct explanation of the difference in melting points of diamond and hydrogen chloride? Tick ✓ **one** box.

A. Strong covalent bonds are broken in diamond but only weak intermolecular forces are broken in hydrogen chloride. ☐

B. The covalent bonds in hydrogen chloride are weaker than those in diamond. ☐

C. Elements have higher melting points than compounds. ☐

④ Which is the correct explanation of the non-conduction of electricity in covalent compounds? Tick ✓ **one** box.

A. The covalent bonds are too strong. ☐

B. The structures have no charged particles free to move. ☐

C. Only transition metals conduct electricity. ☐

Chemistry

Sample response

When describing covalent molecules you should:

- describe covalent bonds forming by sharing pairs of electrons
- know that only non-metal atoms share electrons and form molecules
- be able to work out the number of electrons to be shared by calculating how many electrons are needed to fill the outer shell
- be able to draw dot-and-cross diagrams for H_2, Cl_2, O_2, N_2, HCl, H_2O, NH_3 and CH_4
- explain that a molecular covalent substance has a low melting point because there are weak intermolecular forces (not weak covalent bonds).

Hydrogen, oxygen and nitrogen are made of small molecules.

1 This question will help you draw a dot-and-cross diagram of a nitrogen molecule, N_2, showing the outer shells only.

a Complete 🖉 the sentences.

Nitrogen is in group five so each atom has .... outer electrons. To have a full

outer shell of 8 electrons, each atom must share .... other electrons.

b Add 🖉 the electrons to the shells on the right.
Use dots for one atom and crosses for the other atom.

N   N

c Circle Ⓐ the correct word in the following sentence.

There is a **single** / **double** / **triple** bond in a nitrogen molecule.

2 Draw a dot-and-cross diagram of an oxygen molecule, O_2, showing the outer shells only. Use dots for one atom and crosses for the other atom.

Look at the sample student answer to this question, and list 🖉 the errors on the right.

error 1 ....
....
error 2 ....
....
error 3 ....
....

3 What is the melting point for hydrogen?

Tick ✓ **one** box.

−259°C ☐   40°C ☐   200°C ☐

Hydrogen is a gas at room temperature (20°C). Solid hydrogen must melt and boil below 20°C for it to form a gas at 20°C. Look at each possible answer in turn. Which melting point is consistent with hydrogen being a gas at 20°C?

Your turn!

It is now time to use what you have learned to answer the exam-style question below.

Remember to read the question thoroughly, looking for clues.

Make good use of your knowledge from other areas of chemistry.

Exam-style question

1   Ammonia, NH_3, is made of small molecules.

   1.1   Describe how the atoms are held together in an ammonia molecule.

   ....

   ....

   ....   **(2 marks)**

   > You need to **describe** a bond between a nitrogen atom and a hydrogen atom.
   > These are both non-metal atoms.

   1.2   Draw the dot-and-cross diagram of an ammonia molecule.   **(2 marks)**

   > **Remember** In the molecule each atom should have a full outer shell:
   > • A hydrogen atom is stable with a full outer shell of two electrons.
   > • A nitrogen atom is stable with a full outer shell of eight electrons.

   (2 marks)

   1.3   The boiling point of ammonia is −33 °C. The N–H bonds in ammonia are strong.

   Explain why ammonia has a low boiling point even though the N–H bonds are strong.

   ....

   ....

   ....   **(3 marks)**

   > In this question you are asked to **explain**.
   > – *Describe* the forces between the molecules.
   > – *Explain* why the strength of these forces means that ammonia has a low melting point.

   1.4   Does liquid ammonia conduct electricity? Explain your answer.

   ....

   ....

   ....   **(2 marks)**

   > In this question you are asked to **explain**.
   > – *State* whether ammonia conducts electricity or not.
   > – *Explain* why by talking about the lack of charged particles that are free to move.

Need more practice?

In the exam, questions involving covalent molecules could occur as:
- simple standalone questions
- part of a question on, for example, crude oil
- part of a question about different types of structure and bonding.

Have a go at this exam-style question.

Exam-style question

1 1.1 Draw a dot-and-cross diagram of a molecule of hydrogen, H_2. Use a dot for one atom and a cross for the other atom.

(1 mark)

1.2 Draw a dot-and-cross diagram of a molecule of methane, CH_4, showing the outer shells only. Use dots for the carbon atom and crosses for the hydrogen atoms.

(2 marks)

1.3 Look at the information about **X**, **Y** and **Z**.

	Conductor of electricity?	Melting point (°C)
X	no	1600
Y	yes	650
Z	no	−220

Which substance is made of small molecules and which substance has a giant covalent lattice? Tick **one** box for each type of substance.

Small molecules: X ☐   Y ☐   Z ☐

Giant covalent lattice: X ☐   Y ☐   Z ☐

(2 marks)

Boost your grade

To improve your grade, make sure you learn about the structure and properties of other covalently bonded substances:
- polymers, that have molecules with long chains
- giant covalent substances including carbon in the forms of diamond, graphite and graphene.

You should also know about the structure and properties of metals and of ionic compounds.

How confident do you feel about each of these **skills?** Colour in the bars.

1 How can I describe bonding in substances with only non-metal atoms?

2 How do I draw a dot-and-cross diagram of a molecule?

3 How can I explain the properties of a covalently bonded substance?

④ Rates of reaction

Some reactions are fast and some are slow. The rate of reaction can depend on temperature, pressure (for gases), concentration (for solutions), surface area (for solids) and catalysts. This unit will help you to understand **collision theory** and to explain how different **variables** affect the **rate of reaction**.

In the exam you will be asked to tackle questions such as the one below.

Exam-style question

1 Chips of marble (calcium carbonate) react with dilute hydrochloric acid and produce carbon dioxide gas.

1.1 Plan an experiment to investigate how the rate of this reaction changes when the temperature of the acid is changed. The apparatus that can be used is shown on the left and the results of a similar experiment are shown on the right.

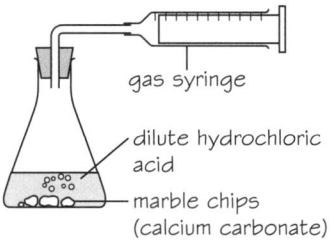
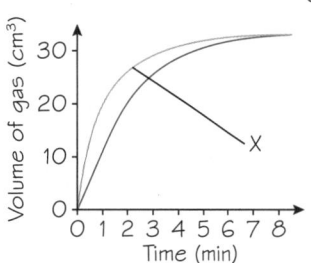

.... (4 marks)

1.2 The similar experiment was carried out at both 30 °C and at 40 °C. The results are shown on the graph above. Explain whether the curve labelled X shows the results at 30 °C or at 40 °C.

.... (2 marks)

1.3 Explain, using collision theory, why the rate of reaction is different at these two temperatures.

.... (3 marks)

You will already have done some work on rates of reaction. Before starting the **skills boosts**, rate your confidence in each area. Colour in the bars.

 1 How can I plan a method to follow a rate of reaction experiment?

 2 How can I get information from a rate of reaction graph?

3 How can I explain why temperature influences the rate of reaction?

Chemistry

For substances to react:
- their particles must collide
- their particles must have enough energy for the collisions to produce a reaction.

A reaction will be faster if there are more particles available to react or if the particles have more energy.

The **rate of reaction** tells us how fast a product is formed (or how fast a reactant is used up).

(1) Zinc, a metal, reacts with hydrochloric acid to form a salt and hydrogen gas.

In an experiment, lumps of zinc are dropped into a beaker of hydrochloric acid. Use ✏ words from the box to say how the following changes will alter the rate of the reaction.

> increases   decreases   does not change

(a) Dilute the hydrochloric acid. The rate of reaction ....

(b) Break the zinc lumps down into powder. The rate of reaction ....

(c) Heat up the hydrochloric acid. The rate of reaction ....

(d) Add a catalyst. The rate of reaction ....

(e) Carry out the reaction in a flask, not a beaker. The rate of reaction ....

(2) The volume of hydrogen gas given off during a reaction between zinc and hydrochloric acid is measured.

Time (minutes)	Volume of gas (cm³)
0	0
1	14
2	27
3	36
4	37
5	50
6	58
7	59
8	60
9	60

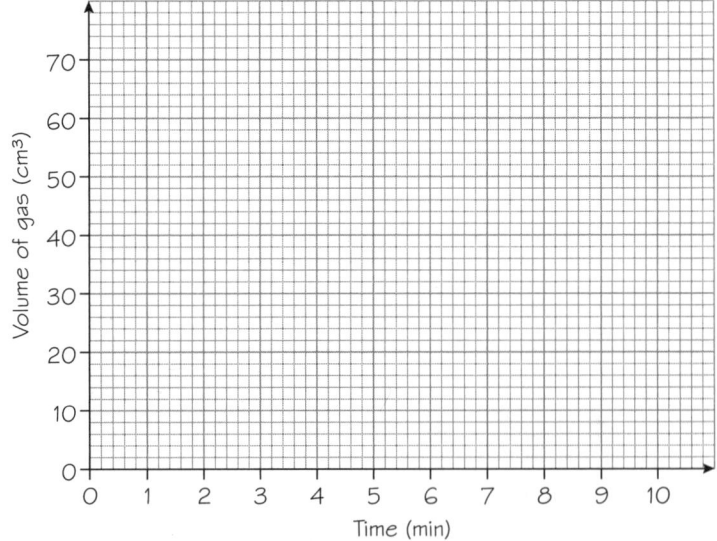

(a) Plot ✏ the results on the grid above, and circle the anomalous point.

An anomalous point is one which does not fit into the pattern.

(b) Draw ✏ a smooth curve of best fit.

A smooth curve of best fit will pass through, or very near, all of the points, except the anomalous point.

(c) Use ✏ evidence from the table to explain when the reaction has finished.

The question says 'use evidence', so you must use some data from the table.

....

....

....

1 How can I plan a method to follow a rate of reaction experiment?

In rate of reaction experiments, only one **variable** must be changed to ensure a fair test.
Reactants should be measured out using appropriate apparatus.

Sodium thiosulfate solution reacts with dilute hydrochloric acid.
A yellow **precipitate** of sulfur forms. The time taken for the sulfur to obscure a cross under the apparatus can be measured.

In an experiment, the concentration of the sodium thiosulfate solution was changed.

The method was as follows.

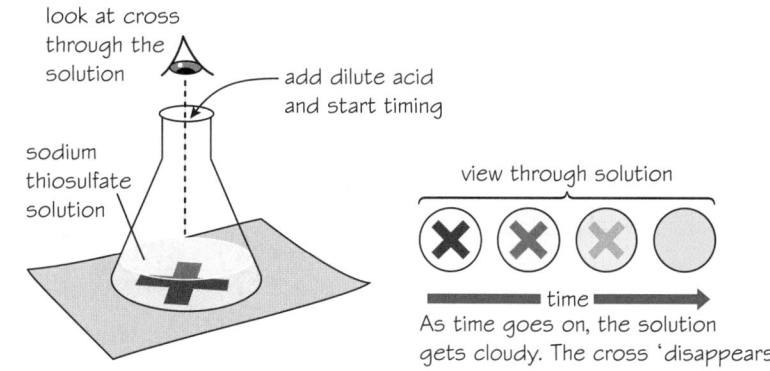

As time goes on, the solution gets cloudy. The cross 'disappears'.

Remember to wear eye protection

i   Measure the sodium thiosulfate solution and water and pour into a flask. Place the flask over a cross.

ii Measure out 5 cm³ acid and add to the flask. Then start the stop clock and swirl.

iii Stop the clock when the cross can no longer be seen.

1   Some things must be kept the same every time the experiment is carried out, to ensure a fair test.

a   Match ✎ each error to the effect it will have on the time recorded.

using a thicker cross when repeating an experiment	the time recorded will be shorter
starting the stop clock after swirling	no effect on the time
using acid of a higher concentration	the time recorded will be longer

b   For this experiment to be fair, the cross must be observed through the same amount of liquid each time. What should be used to measure the liquids accurately? ✎

2   The volumes (in cm³) of sodium thiosulfate and water used in a thiosulfate/acid experiment are shown in the table.

a   Why is water added each time? Tick ✓ your choice.

• to make a solution of sodium thiosulfate ☐

• so the reaction does not get too hot ☐

• so the total volume of the mixture is the same ☐

Experiment number	Sodium thiosulfate	Water
1	5	20
2	10	15
3	15	15
4	20	5
5	25	0

b   In which experiment has an incorrect volume of water been added? ✎

Check the total volume of liquid for each experiment.

2 How can I get information from a rate of reaction graph?

The amount of product formed in a reaction can be measured. If you draw a graph with the time on the x-axis and the amount of product on the y-axis, the slope of the graph is a measure of the rate of reaction. The steeper the slope, the faster the reaction.

1 Hydrogen peroxide decomposes to form water and oxygen.

50 cm³ hydrogen peroxide solution is measured out and allowed to decompose.

The oxygen formed is collected for 8 minutes. The volume of oxygen is shown on the graph.

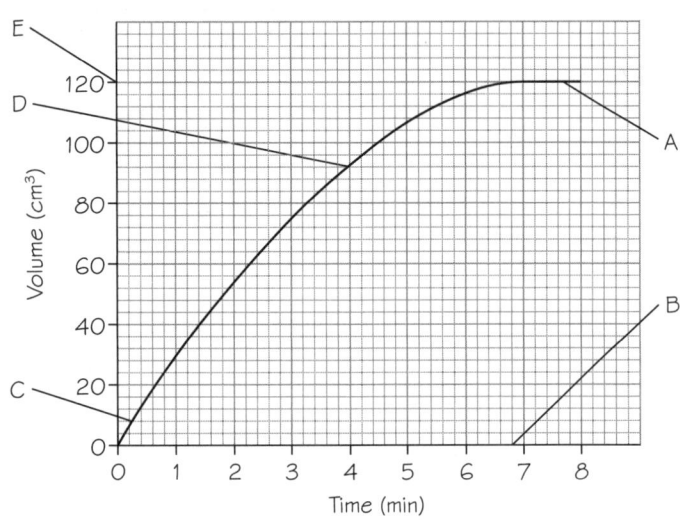
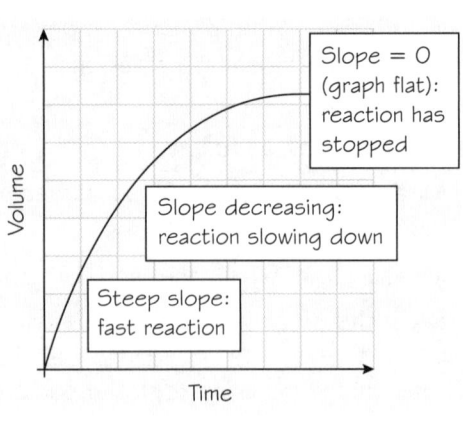

a Look at the graph on the left. Match 🖉 each letter on the graph with the correct statement.

☐ The reaction is fastest here.

☐ The reaction is slowing down here as the hydrogen peroxide is used up.

☐ The reaction has stopped here.

☐ The total volume of oxygen formed is shown here.

☐ The time for the reaction to end is shown here.

b Estimate 🖉 the volume of oxygen formed after 3 minutes. .... cm³

c Complete 🖉 the sentences by choosing the correct word.

i Where the reaction is fastest, the slope of the graph is **greatest** / **smallest**.

ii The reaction slows down because the hydrogen peroxide is **cold** / **used up** / **stronger**.

d Sketch 🖉 on the axes above the curve you would get if 25 cm³ of the hydroxide peroxide solution was diluted with 25 cm³ water and allowed to decompose for 8 minutes.

- Start the curve at 0, 0.
- The number of hydrogen peroxide molecules has been halved. How will this affect the volume of oxygen formed?
- The hydrogen peroxide has been made more dilute. Will this make the reaction faster or slower?

③ How can I explain why temperature influences the rate of reaction?

Particles must have a certain amount of energy to react. If particles with low energy collide, they will not react. Particles with high energy will react when they collide.

The minimum amount of energy the particles must have to react is called the **activation energy**.

Collision theory tells us that a reaction can occur if:

1 the reactant particles collide **and**

2 the particles have enough energy (this is the activation energy).

A reaction will be faster if:

• the particles collide with each other more frequently

• more of the particles have the activation energy.

no reaction   reaction!

low energy collision   high energy collision

① What happens when the temperature is increased? Circle Ⓐ the correct words.

   a The particles will **gain** / **lose** energy.

   b A **lower** / **higher** proportion of the particles will have the activation energy.

   c The particles will move **at the same speed** / **faster**.

   d There will be **more** / **the same number of** collisions every second.

② What effect does increasing the temperature have on the rate of reaction? ✎

....

....

Use your answers to **①** to help you.

③ A catalyst is a substance that increases the rate of a reaction by providing an alternative pathway for the reaction, which has a lower activation energy. Circle Ⓐ the correct words in the sentences below.

> A catalyst provides a pathway of lower activation energy for a reaction so **more** / **fewer**
>
> particles have the activation energy. This makes the reaction **faster** / **slower**.

④ The diagram below is a **reaction profile**. This shows the activation energy the reactants must have in order to react and form the products.

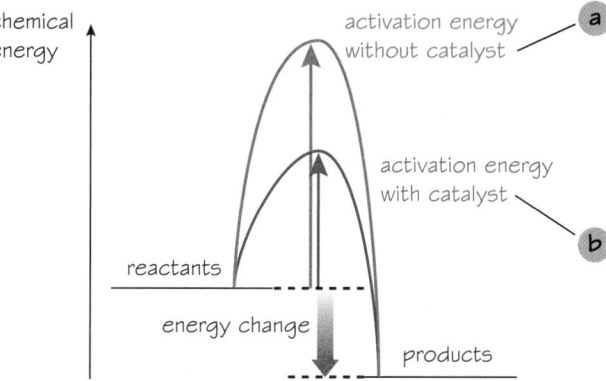

chemical energy

activation energy without catalyst

activation energy with catalyst

reactants

energy change

products

a Without a catalyst, particles need this much energy to react. To increase the rate of reaction, you need to give the particles more energy. How can you do this? ✎

....

b With a catalyst, particles only need this much energy to react. What effect will this have on the rate of reaction? ✎

....

   c Explain ✎ your answer to **b**.

....

....

Chemistry

Sample response

When explaining how the rate of reaction is affected by changing conditions, consider the following points.

1 Increasing the frequency of collisions increases the rate of reaction. You can do this by:
- increasing the concentration of reactants in solution
- increasing the pressure of reacting gases
- increasing the surface area of solid reactants (by making into smaller pieces)
- increasing the temperature.

2 Increasing the number of reactant particles with the activation energy increases the rate of reaction. You can do this by:
- increasing the temperature
- adding a catalyst.

Exam-style question

1 A student investigates how long a reaction takes. The experiment is repeated three more times, adding lumps of three different possible catalysts (P, Q and R) each time.

The results are shown in the table.

Potential catalyst added	Time (s) for reaction to end
none	267
P	34
Q	260
R	43

1.1 Use the data to explain if any of P, Q and R are catalysts for the reaction.

Q is the catalyst for this reaction. I know this because Q has the shortest time.

① Tick ✓ the statements which explain why this is a weak answer.

The answer should refer more clearly to the data in the table.

The answer should mention P, Q and R and say whether each one is a catalyst.

All three of the substances are catalysts.

The reason given for choosing Q does not explain properly why Q is a catalyst.

Q is not a catalyst.

Two of the substances are catalysts.

Catalysts make a reaction **much** faster.

② In another repeat of the experiment, lumps of substance S are found to be a catalyst.

a Give two ways in which the reaction could be made to go even faster than by adding lumps of S.

....

....

b Complete the sentence.

A catalyst works by providing a pathway of .... activation energy so that a .... proportion of the reactant particles now have the activation energy.

Your turn!

It is now time to use what you have learned to answer the exam-style question below. Remember to read the question thoroughly, looking for clues. You may need more space to write your answer, if you do, use paper. Make good use of your knowledge from other areas of chemistry.

Exam-style question

1 Chips of marble (calcium carbonate) react with dilute hydrochloric acid and produce carbon dioxide gas.

1.1 Plan an experiment to investigate how the rate of this reaction changes when the temperature of the acid is changed. The apparatus that can be used is shown on the left and the results of a similar experiment are shown on the right. **(4 marks)**

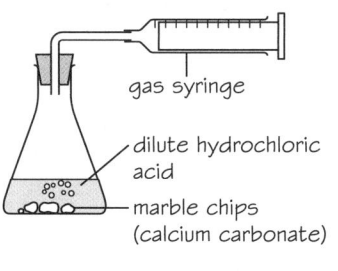

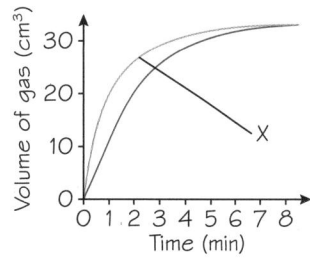

....

....

Describe the readings you would take at room temperature.

....

....

What will you alter and what will you keep the same each time?

....

....

You need to change the temperature. An easy way to do this is by using a water bath that can be set to the different temperatures you need.

1.2 The similar experiment was carried out at both 30 °C and at 40 °C. The results are shown on the graph above. Explain whether the curve labelled X shows the results at 30 °C or at 40 °C. **(2 marks)**

....

....

1.3 Explain, using collision theory, why the rate of reaction is different at these two temperatures. **(3 marks)**

....

....

....

In this question you are asked to **explain**.
• Consider how fast the particles are moving.
• Consider the energy that the particles have. How many have the activation energy?

Chemistry

Need more practice?

In the exam, questions involving rates of reaction could occur as:

• simple standalone questions

• part of a question on, for example, rates and energy changes

• part of a question about a practical experiment.

Have a go at this exam-style question.

Exam-style question

1   A reaction between two gases, X and Y, makes the product P.
    The equation is:

    At a higher pressure, the particles are closer together.

$$X \, (g) + Y \, (g) \rightarrow 2P \, (g)$$

1.1   Explain why an increase in pressure would increase the rate of reaction.

....

....

.... **(3 marks)**

1.2   Consider the rate of production of P. Tick **one** box.

A. The rate of production of P is the same as the rate that X is used up. ☐

B. The rate of production of P is twice as fast as X is used up. ☐

C. The rate of production of P is half as fast as X is used up. ☐   **(1 mark)**

The equation shows that when one molecule of X is used up, two molecules of P are formed.

Boost your grade

To improve your grade, make sure you practise calculating the rate of reaction using the formula:

$$\text{average rate of reaction} = \frac{\text{quantity of product formed (or reactant used up)}}{\text{time taken}}$$

You should try to explain how catalysts increase the rate of reaction using a reaction profile.

How confident do you feel about each of these **skills?** Colour in the bars.

1   **How can I plan a method to follow a rate of reaction experiment?**

2   **How can I get information from a rate of reaction graph?**

3   **How can I explain why temperature influences the rate of reaction?**

(5) Electrolysis

This unit will help you to understand electrolysis. Electrolysis is the breaking down (decomposition) of a compound using electricity.

In the exam you may need to look at a reaction and identify what has been **oxidised** (had oxygen added) and what has been **reduced** (had oxygen removed). You will have to explain how an ionic compound, when melted or dissolved in water, can be electrolysed (broken down using electricity). Also, you should be able to predict the products when a compound is electrolysed.

In the exam you will be asked to tackle questions such as the ones below.

Exam-style questions

1   A solid cube of potassium chloride has electrodes attached to it. It is connected to a circuit to see whether it conducts electricity.

   1.1   Describe how potassium ions and chloride ions are arranged in solid potassium chloride.   **(2 marks)**

   1.2   Explain why solid potassium chloride does **not** conduct electricity.   **(1 mark)**

   1.3   Give two methods by which the solid potassium chloride can be made to conduct electricity.   **(2 marks)**

   1.4   Explain why the methods you gave in **1.3** allow potassium chloride to conduct electricity.   **(1 mark)**

2   **2.1**   Some solid sodium bromide is melted, and the liquid is then electrolysed.

      A silver solid is formed at one electrode and a red-brown liquid is formed at the other electrode.

      Identify the two products.   **(2 marks)**

   2.2   Some solid sodium bromide is dissolved in water, and the solution is then electrolysed.

      A colourless gas is formed at one electrode and an orange solution is formed at the other electrode.

      Identify the two products.   **(2 marks)**

   2.3   When molten lead bromide is electrolysed, the reaction at the cathode is:

$$Pb^{2+} + 2e^- \longrightarrow Pb$$

      Explain why this reaction is described as reduction.   **(1 mark)**

You will already have done some work on electrolysis. Before starting the **skills boosts**, rate your confidence in each area. Colour in the bars.

1 How can I explain oxidation and reduction?	**2** How can I identify the products when a molten substance is electrolysed?	**3** How can I identify the products when an aqueous solution is electrolysed?

Chemistry

Ionic substances can conduct electricity only if the ions are able to move. This happens when the ionic substance is melted or is dissolved in water. When an ionic liquid or solution conducts electricity, it is decomposed as the positive ions are separated from the negative ions. This is called **electrolysis**.

Compounds between metallic elements and non-metallic elements are **ionic**.

In ionic compounds, the metal atoms form **positive ions** and the non-metal atoms form **negative ions**.

(**1**) The diagram shows a lattice of sodium chloride.

   a Label 🖉 a sodium ion and a chloride ion on the diagram.

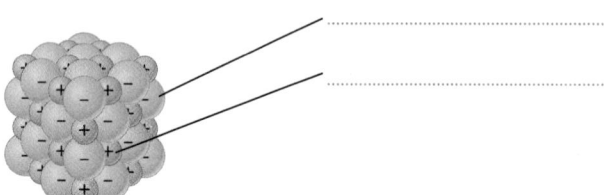

   b Write 🖉 a short paragraph explaining how you worked out which element formed the positive ion and which element formed the negative ion.

Remember You can tell that sodium chloride is an ionic compound because sodium is a metal and chlorine is a non-metal.

....

....

....

(**2**) Sodium chloride dissolves in water. In the solution, the sodium ions and the chloride ions are freed from the solid lattice and mix with the water molecules. The ions are free to move around.

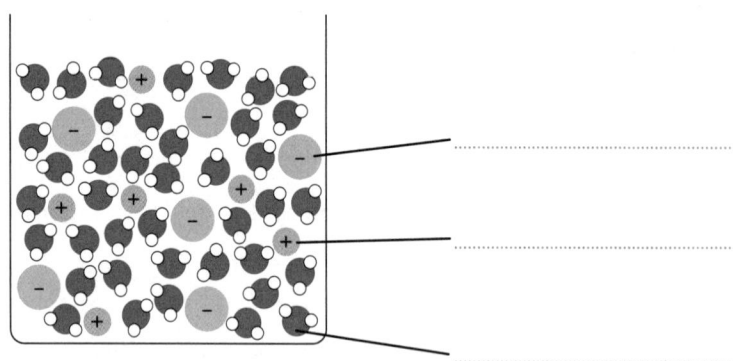

   Label 🖉 a sodium ion, a chloride ion and a water molecule on the diagram.

(**3**) An ionic substance can only conduct electricity when the ions are free to move around.

   Complete 🖉 the table below, showing when ionic compounds can conduct electricity.

Ionic substance	State symbol	Description	Conducts electricity?
solid	s	ions are fixed in a lattice	
melted to a liquid	l	ions are free to move	
dissolved in water	aq	ions are free to move	

1 How can I explain oxidation and reduction?

Oxidation occurs when a substance gains oxygen or loses electrons.

Reduction occurs when a substance loses oxygen or gains electrons.

A substance is oxidised when it reacts with oxygen and bonds with oxygen atoms. If oxygen atoms are removed from a substance, it is reduced.

(1) Some copper is heated in air. Copper oxide is formed.

$$copper + oxygen \rightarrow copper\ oxide$$

Circle Ⓐ the correct words to complete the sentences.

Copper has **gained** / **lost** oxygen atoms in this reaction. Copper has been **oxidised** / **reduced**.

(2) Iron oxide reacts with carbon monoxide as shown in the equation.

$$Fe_2O_3 + 3CO \rightarrow 2Fe + 3CO_2$$

a **i** Identify 🖉 the number of atoms in iron oxide (on the left) and in iron (on the right).

Fe_2O_3 has .... Fe atoms and .... O atoms.

2Fe has .... Fe atoms and .... O atoms.

ii Circle Ⓐ the correct words to complete the sentences.

The iron oxide has **gained** / **lost** oxygen atoms. Iron oxide has been **oxidised** / **reduced**.

b **i** Identify 🖉 the number of atoms in carbon monoxide (on the left) and in carbon dioxide (on the right).

3CO has .... C atoms and .... O atoms.

$3CO_2$ has .... C atoms and .... O atoms.

ii Circle Ⓐ the correct words to complete the sentences.

The carbon monoxide has **gained** / **lost** oxygen atoms. Carbon monoxide has been **oxidised** / **reduced**.

Another way that oxidation and reduction can happen is by gaining or losing electrons.

- When a substance loses electrons, it is oxidised.

- When a substance gains electrons, it is reduced.

(3) **a** $Zn^{2+} + 2e^- \rightarrow Zn$

In this equation, the zinc ions are reduced. Circle Ⓐ the electrons that are gained.

b $2Cl^- \rightarrow Cl_2 + 2e^-$

In this equation, the chloride ions are oxidised. Circle Ⓐ the electrons that are lost.

Remember Oxidation is gaining oxygen atoms and reduction is losing oxygen atoms.

Oxidation **Is L**oss of electrons, **R**eduction **Is G**ain of electrons. Remember this as OILRIG

Chemistry

2 How can I identify the products when a molten substance is electrolysed?

Ionic substances conduct electricity when they are melted or dissolved in water.

When a molten ionic substance is electrolysed, the ions become separated. The substance is decomposed to its elements.

The electrolysis of ionic compounds dissolved in water is discussed on page 93.

Ionic compounds form when a metal reacts with a non-metal:

- the metal always forms a positive ion
- the non-metal always forms a negative ion.

1 Electricity is passed through a molten ionic compound, as shown in the diagram.

a On the diagram label 🖉 the Bunsen burner, the positive electrode and the negative electrode.

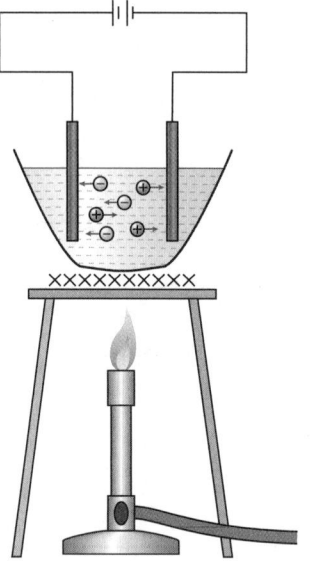

Remember Positive ions are attracted to the negative electrode and negative ions are attracted to the positive electrode.

b Why is the ionic compound heated? Tick ✓ your choice.

i to dry out the compound

ii to speed up the reaction

iii to melt the compound so the ions are free to move

c Potassium iodide contains potassium ions, K^+, and iodide ions, I^-.

In the box below,

i Underline Ⓐ the ion, process and equation at the positive electrode.

ii Circle Ⓐ the ion, process and equation at the negative electrode.

$2I^- \rightarrow I_2 + 2e^-$	iodide ions, I^-	$K^+ + e^- \rightarrow K$
oxidation	reduction	potassium ions, K^+

Remember OILRIG – Oxidation Is Loss of electrons, Reduction Is Gain of electrons

You do not need to learn the equations involving electrons but you should be able to recognise which is oxidation and which is reduction using OILRIG.

d Complete 🖉 the overall equation:

potassium iodide $\rightarrow$ .... + ...

3 How can I identify the products when an aqueous solution is electrolysed?

An aqueous solution contains:
- the positive ions and negative ions from the dissolved substance
- some H^+ ions and OH^- ions from the water.

All these ions need to be considered when you electrolyse an aqueous solution.

(1) Potassium iodide contains K^+ ions and I^- ions. In an aqueous solution of potassium iodide, a few of the water molecules (H_2O) split up to form H^+ ions and OH^- ions.

a Underline (A) the ions present when potassium iodide is dissolved.

| K^+ | I^- | KI | H^+ | OH^- | H_2O |

b If potassium iodide solution is electrolysed, the positive ions are attracted to the negative electrode and the negative ions are attracted to the positive electrode.

Circle (A) the correct words in the sentences below.

i K^+ and H^+ are attracted to the **positive** / **negative** electrode.

ii I^- and OH^- are attracted to the **positive** / **negative** electrode.

When a solution is electrolysed, the product formed at each electrode can be predicted. The product could be from the ionic compound being electrolysed or it could be hydrogen or oxygen from the water.

- At the negative electrode, hydrogen is formed **unless** the metal in the compound is less reactive than hydrogen. If the metal is less reactive than hydrogen then the metal is formed.

  You may need to look at a reactivity series to work this out.

- At the positive electrode, oxygen is formed **unless** a chloride, bromide or iodide compound is electrolysed. If a halide is electrolysed then the halogen is formed.

  'Halide' means chloride, bromide or iodide. 'Halogen' means chlorine, bromine or iodine.

(2) Complete (✎) this sentence.

Potassium is a reactive metal.

When potassium iodide solution is electrolysed, ... is formed at the negative electrode and .... is formed at the positive electrode.

(3) A solution of copper sulfate is electrolysed. Circle (A) the product.

a Copper ions and hydrogen ions are attracted to the negative electrode. Which product forms here?

copper / **hydrogen**

b Sulfate ions and hydroxide ions are attracted to the positive electrode. Which product forms here?

sulfur / **oxygen** / **hydrogen**

c Which of these reactions a or b, is reduction? (✎)

Use OILRIG

Chemistry

Sample response

When describing and predicting electrolysis, remember that:
- ions must be free to move when an ionic substance is electrolysed
- if a molten ionic compound is electrolysed, the elements that make up the compound are formed at the electrodes
- if a solution of an ionic compound is electrolysed, hydrogen or an unreactive metal will form at the negative electrode
- if a solution of an ionic compound is electrolysed, oxygen or a halogen will form at the positive electrode
- oxidation is gaining oxygen or losing electrons
- reduction is losing oxygen or gaining electrons.

Exam-style question

1 Describe an experiment in which molten zinc chloride is electrolysed. Give your observations.

1.1 Write the products that are formed at each electrode and explain whether oxidation or reduction occurs at each electrode.

Here is a sample student answer.

> Zinc chloride powder is placed in a crucible and the apparatus is set up as shown in the diagram.
>
> The electricity is turned on. At the negative electrode a pale green gas is seen.
>
> At the positive electrode a liquid metal is seen.
>
> At the negative electrode, chloride is formed. At the positive electrode, zinc is formed.
>
> Zinc ions are Zn^{2+} so they must gain electrons to form zinc atoms.
>
> Chloride ions are Cl^- so they must lose electrons to form chlorine molecules.

① What is missing from the circuit? ✎

See page 92 for a complete circuit.

....

② The safety symbol on zinc chloride is shown to the right.
Explain ✎ a safety precaution that should be taken when adding the zinc chloride powder.

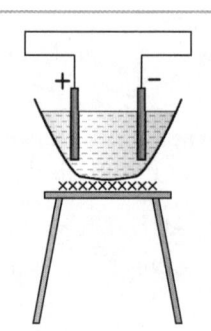

This safety symbol is also used on some acids.

....

③ What should have been done to the zinc chloride to allow it to conduct electricity? Tick ✓ the correct box.

The ions must be free to move.

heat it until it melts ☐   grind it to a powder ☐   dissolve it with ethanol ☐

④ The colour of the liquid metal is missing. Give ✎ the correct colour.

....

Copper is red-brown and gold is yellow, but what colour are most metals?

⑤ The element formed at the negative electrode has been named incorrectly.
Write ✎ the correct name of the element.

....

⑥ The question asks whether oxidation or reduction occurs. Circle Ⓐ the correct answers.

> The zinc ions are **oxidised** / **reduced**. The chloride ions are **oxidised** / **reduced**.

Your turn!

It is now time to use what you have learned to answer the exam-style questions below.
Remember to read the question thoroughly, looking for clues.
Make good use of your knowledge from other areas of chemistry.

Exam-style questions

1   A solid cube of potassium chloride has electrodes attached to it. It is connected to a circuit to see whether it conducts electricity.

   1.1   Describe how potassium ions and chloride ions are arranged in solid potassium chloride.

   ....

   ....   **(2 marks)**

   1.2   Explain why solid potassium chloride does **not** conduct electricity.   Can the ions move?

   ....   **(1 mark)**

   1.3   Give two methods by which the solid potassium chloride can be made to conduct electricity.

   ....

   ....   **(2 marks)**

   1.4   Explain why the methods you gave in **1.3** allow potassium chloride to conduct electricity.

   ....   **(1 mark)**

2   2.1   Some solid sodium bromide is melted, and the liquid is then electrolysed.

        A silver liquid is formed at one electrode and a red-brown liquid is formed at the other electrode.

        Identify the two products.   The sodium bromide is melted, so the only possible products are the elements from which sodium bromide is made.

   ....

   ....   **(2 marks)**

   2.2   Some solid sodium bromide is dissolved in water, and the solution is then electrolysed.

        A colourless gas is formed at one electrode and an orange solution is formed at the other electrode.

        Identify the two products.   The information on page 93 will help you to answer this question.

   ....

   ....   **(2 marks)**

   2.3   When molten lead bromide is electrolysed, the reaction at the cathode is:

$$Pb^{2+} + 2e^- \rightarrow Pb$$

        Explain why this reaction is described as reduction.   Use **OILRIG**

   ....   **(1 mark)**

Need more practice?

In the exam, questions involving electrolysis could occur as:
- simple standalone questions
- part of a question on, for example, ionic compounds
- part of a question about a practical experiment.

Have a go at this exam-style question.

Exam-style question

1   Aluminium oxide can be made by reacting aluminium with oxygen:

$$aluminium + oxygen \longrightarrow aluminium\ oxide$$

   1.1   Explain whether the aluminium is oxidised or reduced in this reaction.

   ....

   ....

   (2 marks)

   1.2   Aluminium oxide is insoluble in water. How can solid   You can't dissolve aluminium oxide
            aluminium oxide be made to conduct electricity?   in water. What else could you do?

   ....

   (1 mark)

   1.3   Aluminium oxide contains Al^{3+} ions and O^{2-} ions.

            Complete the equations showing the reactions that occur at the electrodes
            when aluminium oxide is electrolysed.
            Say whether oxidation or reduction has happened.   **(4 marks)**

            $Al^{3+} +$ .... $e^- \longrightarrow Al$ this is **oxidation / reduction**   How many electrons
                                                                                                                         are needed to balance
            $2O^{2-} \longrightarrow$ .... $+ 4e^-$ this is **oxidation / reduction**   the 3+ charge on an
                                                                                                                         aluminium ion?
   You need to give the correct formula for oxygen gas and then check
   you have balanced the number of oxygen atoms in this equation.

Boost your grade

To improve your grade, make sure you practise writing or completing half-equations showing positive
ions gaining electrons (reduction) and negative ions losing electrons (oxidation).
You could learn how aluminium and copper are produced industrially using electrolysis.
You could also look at the reactivity series, as you need to know if metals are more reactive or less
reactive than hydrogen.

How confident do you feel about each of these **skills?** Colour in the bars.

1 How can I explain oxidation and reduction?

2 How can I identify the products when a molten substance is electrolysed?

3 How can I identify the products when an aqueous solution is electrolysed?

⑥ Calculations

This unit will help you to understand how to calculate the masses of atoms, and of reactants and products, in a reaction.

In the exam you may need to calculate:

• the average mass of the atoms of an element

• the mass of a product of a reaction when you are told the masses of the reactants

• the mass of substance needed to make up a solution.

In the exam you will be asked to tackle questions such as the ones below.

Exam-style questions

1 Magnesium has three isotopes, shown in the table.

Isotope	Percentage abundance
magnesium-24	79.0
magnesium-25	10.0
magnesium-26	11.0

  1.1 Give the similarity and the difference between atoms of these three isotopes.

                                                                    (3 marks)

  1.2 Calculate the relative atomic mass of magnesium. Give your answer to 1 decimal place.   **(3 marks)**

2 Magnesium reacts with dilute hydrochloric acid to form magnesium chloride and a gas.

  2.1 Complete the symbol equation for the reaction.

      $Mg + 2HCl \longrightarrow MgCl_2 +$ ....   **(1 mark)**

  2.2 Explain why the mass of the flask containing the reactants decreases.   **(1 mark)**

  2.3 Calculate the mass of magnesium chloride that can be formed if 12.0 g of magnesium are added to an excess of acid.

      Relative atomic masses (A_r): Mg = 24, Cl = 35.5   **(3 marks)**

3 A solution of dilute hydrochloric acid was made. 73.0 g of HCl were dissolved to make 2.00 dm³ solution.

Calculate the concentration of the acid in g/dm³.

                         concentration = ... g/dm³   **(2 marks)**

You will already have done some work on calculations. Before starting the **skills boosts**, rate your confidence in carrying out these calculations. Colour in the bars.

1 How can I calculate the relative atomic mass?

2 How can I calculate the mass of a product formed in a reaction?

3 How can I calculate the concentration of a solution?

Chemistry

The **relative atomic mass** shows how the masses of the atoms compare to each other. The lightest atom, hydrogen, has a relative atomic mass of 1 and the most common atom of carbon has a mass of exactly 12.

The **relative formula mass** of a substance is found by adding the relative atomic masses of all the atoms in the formula.

In the exam you will be given a periodic table.

Each element will be shown like this:

```
19 ———— relative atomic mass
F ———— atomic symbol
fluorine ———— name
9 ———— atomic (proton) number
```

(1) The element fluorine is in period 2 and group 7.

   a How many protons are there in an atom of fluorine?

   Circle (A) the correct number.

   | 2 | 7 | 9 | 19 |

   This is the proton number.

   b Fluorine exists as a gas. Molecules of fluorine gas have the formula F_2.

   Calculate the relative formula mass of a fluorine molecule.

   relative formula mass = 2 × .... = ...

   > **Remember** All gaseous elements exist as molecules with two atoms
   > (e.g. O_2, Cl_2, N_2) except the group 0 gases, which exist as separate atoms.

(2) **a** How many atoms of each element are there in the formula Na_2HPO_4?

   Add the numbers round the formula.   Where there is no number in the formula, this means 1.

   $$Na_2HPO_4$$

   oxygen, O = ... atom(s)
   phosphorus, P = ... atom(s)
   sodium, Na = ... atom(s)
   hydrogen, H = ... atom(s)

   b Calculate the relative formula mass of Na_2HPO_4.

   Relative atomic masses (A_r): H = 1, O = 16, Na = 23, P = 31

   relative formula mass = (23 × ....) + (1 × ....) + (31 × ....) + (16 × ....) = ...

(3) **a** How many atoms of each element are there in the formula $Ca(NO_3)_2$?

   Add the numbers round the formula.

   The '2' outside the brackets means there is two times everything inside the brackets

   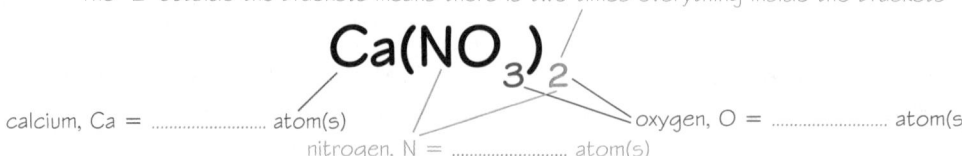

   $$Ca(NO_3)_2$$

   calcium, Ca = ... atom(s)
   nitrogen, N = ... atom(s)
   oxygen, O = ... atom(s)

   b Calculate the relative formula mass of $Ca(NO_3)_2$.

   Relative atomic masses (A_r): N = 14, O = 16, Ca = 40

   relative mass = ...

How can I calculate the relative atomic mass?

All of the atoms of a particular element have the same number of protons. This is the atomic number of the element and it is how the elements are ordered in the periodic table. However, the number of neutrons in the atoms of a particular element can vary, so the mass of atoms of an element can also vary. The **relative atomic mass** is the average relative mass of an atom, taking into account the abundance of different isotopes.

1 There are two isotopes of chlorine, ^{35}Cl and ^{37}Cl.

Remember Isotopes are atoms of an element with different numbers of neutrons.

The percentage abundance of each isotope is $^{35}Cl = 75\%$ and $^{37}Cl = 25\%$.

Remember % means 'out of 100', so 75% = 75 out of 100.

Write ✏ the masses of ^{35}Cl and ^{37}Cl into the equation below to calculate the relative atomic mass of chlorine.

relative atomic mass = $\left(\text{..................} \times \dfrac{75}{100}\right) + \left(\text{..................} \times \dfrac{25}{100}\right) = \text{..................}$

2 Lithium has two isotopes.

Isotope	Percentage abundance
lithium-6	7.6
lithium-7	92.4

Write ✏ the percentages into the equation below to calculate the relative atomic mass of lithium. Give your answer to 1 decimal place.

relative atomic mass = $\left(6 \times \dfrac{\boxed{}}{100}\right) + \left(7 \times \dfrac{\boxed{}}{100}\right) = \text{..................}$

Remember 'Give your answer to 1 decimal place' means you should have **one digit** after the decimal point. To round to one decimal place look at the digit in the 2nd decimal place. If it is 5 or more, round up.

3 Silicon has three isotopes: $^{28}Si = 92.2\%$, $^{29}Si = 4.7\%$, $^{30}Si = 3.1\%$.

Calculate ✏ the relative atomic mass of silicon. Give your answer to 1 decimal place.

....

....

....

4 Bromine has two isotopes, bromine-79 and bromine-81.

The relative atomic mass of bromine is 79.9.

a Calculate ✏ the relative formula mass of bromine, Br_2.

....

....

b Explain ✏ which of the two isotopes of bromine is most common.

....

....

Chemistry

② How can I calculate the mass of a product formed in a reaction?

A chemical equation tells us the numbers of each atom or molecule that react or are formed in a reaction. The relative formula masses can be used to calculate the masses of each substance in the reaction.

Exam-style question

1 Calcium reacts with fluorine to form calcium fluoride: $Ca + F_2 \rightarrow CaF_2$

   1.1 Calculate the mass of calcium fluoride that can be made from 80 g of calcium.

      Relative atomic masses (A_r): F = 19, Ca = 40; relative formula mass: CaF_2 = 78

① Use the numbers from the equation to find the ratio of reacting masses. Fill in the gaps. 🖉

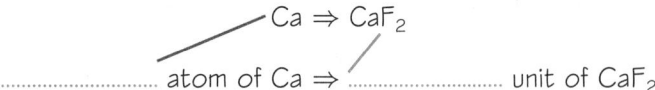

      Ca ⇒ CaF₂

.... atom of Ca ⇒ .... unit of CaF₂

Remember Where there is no number, this means 1.

② Use the information from ① to calculate 🖉 the mass of CaF₂ formed from 80 g Ca.

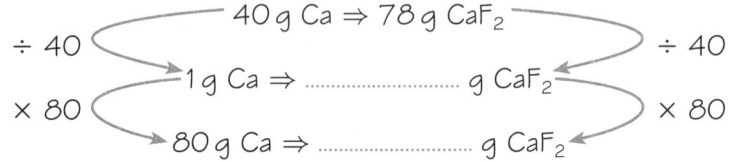

÷ 40 ⎧ 40 g Ca ⇒ 78 g CaF₂ ⎫ ÷ 40

× 80 ⎨ 1 g Ca ⇒ .... g CaF₂ ⎬ × 80

     80 g Ca ⇒ .... g CaF₂

1 Divide by 40 to get 1 g of Ca.
2 Multiply by 80 to get the mass of calcium used in the question.
You need to do the same to the mass of CaF₂ to keep the proportions the same.

Exam-style question

2 Sodium carbonate reacts with hydrochloric acid: $Na_2CO_3 + 2HCl \rightarrow 2NaCl + H_2O + CO_2$

   2.1 How much sodium chloride can be made from 10.6 g of sodium carbonate?

      Relative atomic masses (A_r): C = 12, O = 16, Na = 23, Cl = 35.5

③ Calculate 🖉 the relative formula masses of sodium carbonate and sodium chloride.

  a Na_2CO_3 = (23 × ....) + (12 × ....) + (16 × ....) = 106

  b NaCl = 23 + ... = ...

④ Use the balancing numbers from the equation to calculate 🖉 the ratio of reacting masses.

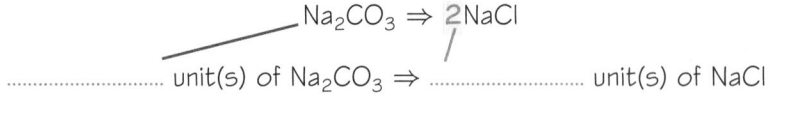

      Na₂CO₃ ⇒ 2NaCl

.... unit(s) of Na₂CO₃ ⇒ .... unit(s) of NaCl

You can use the relative formula masses from ③.

     106 g Na₂CO₃ ⇒ 2 × .... = 117 g NaCl

⑤ Use the information from ④ to calculate 🖉 the mass of NaCl formed from 10.6 g of Na₂CO₃.

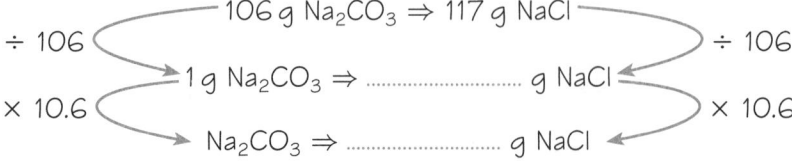

÷ 106 ⎧ 106 g Na₂CO₃ ⇒ 117 g NaCl ⎫ ÷ 106

× 10.6 ⎨ 1 g Na₂CO₃ ⇒ .... g NaCl ⎬ × 10.6

    Na₂CO₃ ⇒ .... g NaCl

③ How can I calculate the concentration of a solution?

Many substances used in reactions are solutions. The concentration of a solution is the mass of the dissolved substance divided by the volume of the solution formed. The units are g/dm^3.

① 36 g of sodium chloride are dissolved in water and the volume of the solution formed is $3\,dm^3$.

Calculate the concentration of the sodium chloride solution.

$$\text{concentration} = \frac{\text{mass in g}}{\text{volume in dm}^3} = \frac{\boxed{}}{\boxed{}} = \text{.....................} \; g/dm^3$$

Remember $\text{concentration (g/dm}^3) = \dfrac{\text{mass of solute (in g)}}{\text{volume of solution formed (in dm}^3)}$

Note how the unit 'g/dm^3' matches the equation, with a mass (in g) divided by the volume (in dm^3).

② 16 g of potassium nitrate are dissolved in water and the volume of the solution formed is $400\,cm^3$.

Calculate the concentration of the potassium nitrate solution.

$$400\,cm^3 = \frac{\boxed{}}{\boxed{}} = \text{.....................} \; dm^3$$

Remember The volume must be in dm^3. $1\,dm^3 = 1000\,cm^3$. To change cm^3 to dm^3, divide by 1000.

$$\text{concentration} = \frac{16\,g}{\boxed{}}$$

$$= \text{.....................} \; g/dm^3$$

③ Calculate the mass of copper sulfate in $2.5\,dm^3$ of copper sulfate solution of concentration $18.6\,g/dm^3$.

This calculation asks for the mass of copper sulfate. The equation can be rearranged to give:

$\text{mass} = 18.6 \times 2.5$ mass of solute (in g) = concentration (in g/dm^3) × volume (in dm^3)

$$= \text{.....................} \; g$$

④ Calculate the mass of copper nitrate in $250\,cm^3$ of copper nitrate solution of concentration $1.60\,g/dm^3$.

$$250\,cm^3 = \frac{\boxed{}}{\boxed{}} = \text{.....................} \; dm^3$$

Remember The volume must be in dm^3.

$\text{mass} = 1.60 \times \text{.....................}$

$$= \text{.....................} \; g$$

⑤ Calculate the mass of sodium chloride in $500\,cm^3$ of a solution of concentration $32.4\,g/dm^3$.

Chemistry

Sample response

Remember the three types of calculation:

1 To calculate the relative atomic mass from isotopes, use mass of isotope × % abundance.

2 To calculate the mass of a product, knowing the mass of a reactant:
  i   calculate the relative formula masses of the two substances
  ii use the numbers from the equation to calculate the ratio of reacting masses
  iii use the information from step ii to find the mass required in the question.

3 To calculate the concentration, use concentration $= \dfrac{\text{mass}}{\text{volume}}$

Look at the sample student answers to the questions below.

Exam-style question

1   Boron has two isotopes, boron-10 (19.9%) and boron-11 (80.1%).

  1.1   Calculate the relative atomic mass of boron. Give your answer to one decimal place.

    relative atomic mass = (10 × 19.9) + (11 × 80.1) = 1080

① **a**   What has the student forgotten to divide by?   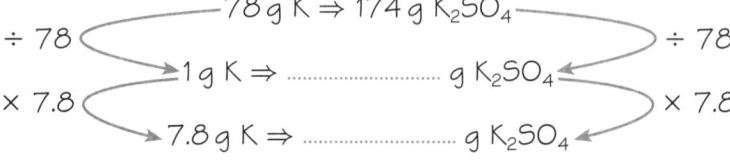 ....   Look at page 99.

  b   Multiply the figures again and divide by the correct number. Write ✎ the number shown on your calculator.

$$\left(10 \times \frac{19.9}{\rule{2cm}{0.4pt}}\right) + \left(11 \times \frac{80.1}{\rule{2cm}{0.4pt}}\right) = \rule{3cm}{0.4pt}$$

② Round ✎ your answer to ① **b** to one decimal place. ....

Exam-style question

2   Potassium reacts with sulfuric acid: $2K + H_2SO_4 \rightarrow K_2SO_4 + H_2$

  2.1   Calculate the mass of K_2SO_4 that can be made from 7.8 g of potassium.

    Relative atomic masses (A_r): O = 16, S = 32, K = 39

    K = 39 K_2SO_4 = (39 × 2) + (32 × 4) + (16 × 4) = 270

③ The actual relative formula mass of K_2SO_4 is 174.

  Circle Ⓐ the part of the student's calculation that is incorrect.

④ Use the numbers from the equation to calculate the ratio of reacting masses. Fill in the gaps. ✎

.... K ⇒ K_2SO_4

.... × 39 g = ... g K ⇒ 174 g K_2SO_4

⑤ Use the information from ④ to calculate ✎ the mass of K_2SO_4 formed from 7.8 g of K.

```
                 ┌── 78 g K ⇒ 174 g K₂SO₄ ──┐
  ÷ 78 ⟨                                      ⟩ ÷ 78
        └→ 1 g K ⇒ ............ g K₂SO₄ ←┘
  × 7.8 ⟨                                      ⟩ × 7.8
        └→ 7.8 g K ⇒ ............ g K₂SO₄ ←┘
```

Your turn!

It is now time to use what you have learned to answer the exam-style questions below.

Remember to read the questions thoroughly, looking for clues.

Make good use of your knowledge from other areas of chemistry.

Exam-style questions

1 Magnesium has three isotopes, shown in the table.

Isotope	Percentage abundance
magnesium-24	79.0
magnesium-25	10.0
magnesium-26	11.0

Mention electrons, protons and neutrons.

1.1 Give the similarity and the difference between atoms of these three isotopes.

....

....

.... (3 marks)

1.2 Calculate the relative atomic mass of magnesium. Give your answer to 1 decimal place.

....

.... (3 marks)

2 Magnesium reacts with dilute hydrochloric acid to form magnesium chloride and a gas.

2.1 Complete the symbol equation for the reaction.

Count all of the atoms on the left. Then see which have not already been used on the right.

$$Mg + 2HCl \rightarrow MgCl_2 + \text{............................}$$

(1 mark)

2.2 Explain why the mass of the flask containing the reactants decreases.

Does one of the products escape? Which one? Why does it escape?

.... (1 mark)

2.3 Calculate the mass of magnesium chloride that can be formed if 12.0 g of magnesium are added to an excess of acid.

Relative atomic masses (A_r): Mg = 24, Cl = 35.5

There are three steps here:
1 Calculate the relative formula masses.
2 Find the ratio of reacting masses.
3 Calculate the reacting mass.
Look back at page 100 if necessary.

....

....

.... (3 marks)

3 A solution of dilute hydrochloric acid was made. 73.0 g of HCl were dissolved to make 2.00 dm³ solution.

Calculate the concentration of the acid in g/dm³.

$$\text{concentration} = \frac{\text{mass}}{\text{volume}}$$

concentration = ... g/dm³ (2 marks)

Need more practice?

In the exam, questions involving calculations could occur as simple standalone questions, part of a question on any chemical reaction or part of a question about a practical experiment.

Have a go at these exam-style questions.

1   Neon has three isotopes.

Isotope	Percentage abundance
neon-20	90.48
neon-21	0.27
neon-22	9.25

    1.1 Calculate the relative atomic mass of neon.

        Give your answer to 2 decimal places.

(3 marks)

2   Sodium hydroxide reacts with sulfuric acid: $2NaOH + H_2SO_4 \rightarrow Na_2SO_4 + 2H_2O$

    Relative atomic masses (A_r): H = 1, O = 16, Na = 23, S = 32

    2.1 Calculate the mass of Na_2SO_4 that can be formed from 16.0 g NaOH.

(4 marks)

3   In the reaction in **2**, 490 g of pure sulfuric acid was dissolved carefully to make 10 dm³ of dilute sulfuric acid.

    3.1 Calculate the concentration of the acid in g/dm³.

$$concentration = \frac{mass}{volume}$$

(2 marks)

To improve your grade, you can try other calculations. For example:
* calculate the number of protons, neutrons and electrons in an atom
* calculate R_f values in chromatography
* calculate the rate of reaction.

How confident do you feel about each of these **skills?** Colour in the bars.

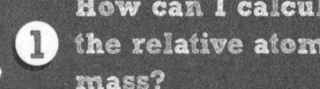

1 How can I calculate the relative atomic mass?

2 How can I calculate the mass of a product formed in a reaction?

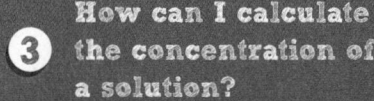

3 How can I calculate the concentration of a solution?

⑦ Answering extended response questions

This unit will help you to understand what is needed for extended response questions. This unit will also help you to plan and write the answer concisely.

In the exam you will need to write extended answers. These answers must be written in continuous prose. They must be logical, with your points arranged in a sensible order. Sometimes, these answers may require you to link together knowledge and understanding from different topics.

In the exam you will be asked to tackle questions such as the one below.

Exam-style question

1   A student wants to investigate which of three substances acts as a catalyst.

Hydrogen peroxide decomposes to form water and oxygen gas.

$$\text{hydrogen peroxide} \rightarrow \text{water} + \text{oxygen}$$

The student is provided with normal laboratory apparatus, and

- hydrogen peroxide solution
- three powders, **P**, **Q** and **R**

1.1   Plan an experiment to investigate which of the powders **P**, **Q** and **R** are catalysts for the decomposition of hydrogen peroxide.

- Name the apparatus used.
- Comment on how to carry out a fair test.
- Explain how you would analyse the results to see which of **P**, **Q** and **R** are catalysts.

.... **(6 marks)**

You will already have written some answers to extended response questions. Before starting the **skills boosts**, rate your confidence in your ability to understand, plan and use the correct amount of detail when answering an extended response question. Colour in 🖊 the bars.

1 How do I know what the question is asking me to do?

2 How do I plan my answer?

3 How do I choose the right detail to answer the question concisely?

An extended response question will sometimes contain a lot of information that can help you. You should use all of the information to find out what you must write.

The **command word**, such as *evaluate*, *explain*, *calculate*, *compare* or *plan*, tells you what type of answer to write.

(1) Match 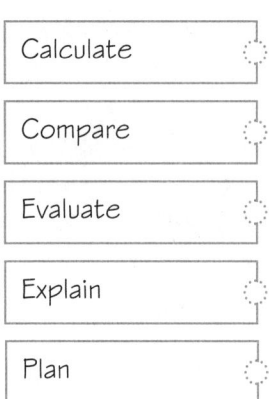 each command word with its description.

Calculate	Use information given in the question and your own knowledge and understanding to consider evidence for and against.
Compare	Give the scientific reasons for something.
Evaluate	Write a method for an experiment.
Explain	Describe the similarities and differences between things.
Plan	Use numbers given in the question to work out the answer.

The key words in the descriptions are underlined.

Atoms contain electrons in shells, and protons and neutrons in a nucleus.

The diagram shows three isotopes of hydrogen.

 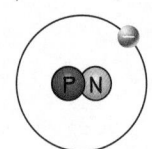 

Compare these isotopes of hydrogen.

This command word means that you should give similarities and differences.

(2) (a) Write the command word for the question above. ....

(b) What are the similarities? Complete the sentences.

i The isotopes have the same number of ....

ii The isotopes have the same number of ....

(c) What are the differences? Complete the sentence.

The isotopes have different numbers of ....

(d) Now put the information from (b) and (c), together with the position of the sub-atomic particles, into a complete answer to the question.

....

....

....

....

How do I know what the question is asking me to do?

You can work out what the question is asking you to do by:
- looking at the command word, e.g. describe, explain, compare
- breaking the question stem into sections to decide which aspects of the topic are being asked about.

Here is some information about ionic and covalent compounds.

> Compounds can have ionic bonding or covalent bonding.
>
> When metal and non-metal atoms react, an ionic compound forms. Ionic compounds consist of a giant ionic lattice of positive metal ions and negative non-metal ions held together by strong ionic bonds.
>
> When non-metal atoms bond together, a covalent compound forms. Some covalent compounds consist of a giant covalent lattice of atoms held together by strong covalent bonds. Other covalent compounds consist of small molecules with only weak intermolecular forces attracting the molecules.

Exam-style question

1   Carbon dioxide melts at −57 °C, silicon dioxide at 1710 °C and sodium oxide at 1132 °C.

   1.1   Explain the melting points of these compounds, referring to their structure and bonding.

Step 1: Think about the question.

① Write 🖉 the command word in this question. ....

② Two scientific ideas are mentioned in the question for you to use in your explanation of the melting points. Circle Ⓐ the two ideas in the question.

> The boxed text contains information you need to answer the question.

Step 2: Think about the information.

③ The text describes three types of bonds or forces. Circle Ⓐ these in **red**.

④ The text describes three types of structure. Highlight 🖉 these types.

Step 3: Think about the answer.

⑤ Your answer needs to link the information you have to the question that is asked.

   Assemble the information by completing 🖉 these sentences.

   ⓐ Sodium oxide is ionic because it is made from a .... and a

   .... element.

   ⓑ Ionic compounds have structures called a giant .... and have high melting

   points because the bonds you need to break are ....

   ⓒ Carbon dioxide and silicon dioxide are covalent because they are made from just

   .... elements.

   ⓓ Carbon dioxide has a low melting point because there are only weak ....

   .... between the molecules.

   ⓔ Silicon dioxide has a high melting point because it has a giant ....

   and there are .... holding the atoms together.

Chemistry

2   How do I plan my answer?

You can plan your answer by:
- thinking through the topic as a whole
- deciding which parts are relevant to the question
- structuring your answer by putting your points in a logical order.

Here is some information about the use of petrol and hydrogen in cars.

> Petrol is obtained in large amounts by the fractional distillation of crude oil. When petrol is burned, carbon dioxide and water are released. Sulfur dioxide, nitrogen oxides and particulates may also be released.
>
> Hydrogen is produced from water by electrolysis. There are only a few places where a driver can get hydrogen. Hydrogen is a gas and has to be stored at high pressure in heavy tanks. When hydrogen is burned, the only product is water.

Exam-style question

1   Both petrol and hydrogen can be used as fuel for cars.

   1.1   Evaluate the use of petrol and hydrogen in cars.

(1) Write 🖉 the command word in this question. ....

(2) The information you are provided with, and the knowledge you have been taught, should be used to answer this question. The advantages and disadvantages of each fuel should be weighed up.

   (a) Highlight 🖉 the advantages of using petrol as a fuel in one colour.

   (b) Highlight 🖉 the disadvantages of using petrol in another colour.

   (c) Circle Ⓐ the advantages of using hydrogen in **red**.

   (d) Circle Ⓐ the disadvantages of using hydrogen in blue.

(3) Now plan 🖉 your answer on paper, structuring your points in a logical order, looking first at the advantages and then the disadvantages of each fuel.

Advantages of petrol
Disadvantages of petrol
Advantages of hydrogen
Disadvantages of hydrogen

How available is the supply? Why is it an advantage being a liquid?

Are the burning products pollutants? Is petrol a renewable resource?

Is the water supply limited? Is the burning product a pollutant?

Can drivers easily get hydrogen? What is the energy supply for electrolysis?

(4) On paper, write 🖉 out your answer in full.

3 How do I choose the right detail to answer the question concisely?

You can get the right amount of detail in your answer by:

- selecting which parts of the topic answer this question, rather than attempting to write everything you know about the whole topic
- referring back to the command word to see the style you should use in your answer.

Exam-style question

1 A life-cycle assessment (LCA) has been carried out on two types of shopping bag. A table of results, adapted from the study, shows the energy consumption and waste mass formed when manufacturing the bags.

Bag	Energy used in MJ	Waste mass in g
single-use plastic bag	22	420
reusable cotton bag	40	1800

1.1 Explain, using **just** this information, why the claim 'cotton bags are better for the environment than plastic bags' is misleading.

The command word in this question is **explain**.

① This is a question in a science exam, so you need to give scientific reasons. What does the data in the table give you information about? ✏

....

....

Analyse the question to make sure you answer all of the parts.

② Highlight ✏ in one colour the part of the question asking you about the positive aspects of using a cotton bag and highlight ✏ the negative aspects in another colour.

Include all of the useful information.

③ What are the two factors mentioned in the table that you should use? ✏

.... and ....

Do not give unnecessary information.

④ Circle Ⓐ the part of the question that tells you to use only the information given to you.

⑤ Use this student plan to help you to write an answer. Write ✏ your answer on a separate piece of paper.

> *Analysis of first factor:*
> - *Compare the energy used in making a cotton bag or a plastic bag.*
>
> *Analysis of second factor:*
> - *Compare the waste mass formed in making a cotton bag or a plastic bag.*
>
> *Bring your analyses together and draw a conclusion:*
> - *How many times does a cotton bag have to be used to use less energy and make less waste per use than a single-use plastic bag?*

Chemistry

Sample response

Remember in extended response questions to:

- look at the command word, to see what you have to do and what style to use
- break the question stem into sections to decide which aspects to mention
- think through the topic as a whole and select the relevant parts to answer the question
- structure your points into a logical order.

Exam-style question

1 Small pieces of lithium, sodium and potassium are dropped into separate troughs of water.

1.1 Describe the observations that you could make in these reactions and explain how these observations can be used to place the metals in order of reactivity.

1 **a** Write ✐ the **two** command words for this question. .... and ....

b Underline Ⓐ the three metals that must be mentioned in the answer.

c Circle Ⓐ the two aspects of the question that must be answered.

2 Now analyse one student's answer. You will need to write ✐ your responses on a separate piece of paper.

a Comment on the correctness and relevance of this part of the answer:

> Lithium is stored in a bottle in oil and is removed with tweezers. Safety glasses must be used. The oil is cleaned off and a small piece of lithium is cut off. This piece is dropped into the water.

b Comment on whether this part of the answer is a good description of what you would observe.

> The lithium moves around on the surface of the water and disappears. It fizzes slowly.

c Comment on whether giving observations for lithium, then for sodium and then for potassium is a good idea.

> When the experiment is repeated with sodium, the sodium moves around faster on the surface of the water. It fizzes quickly and disappears. When the experiment is repeated with potassium, the potassium moves even faster on the surface. The fizzing is very fast and the metal disappears very quickly.

d An observation has been missed in the final part of the answer. Give the missing observation.

e To complete the answer, the order of reactivity must be given. List the metals in the correct order.

The order of reactivity isn't given in the question. It is information you need to provide.

Your turn!

It is now time to use what you have learned to answer the exam-style question below.

Remember to read the question thoroughly, looking for clues.

Make good use of your knowledge from other areas of chemistry.

Exam-style question

1   A student wants to investigate which of three substances acts as a catalyst.

Hydrogen peroxide decomposes to form water and oxygen gas.

$$\text{hydrogen peroxide} \rightarrow \text{water} + \text{oxygen}$$

The student is provided with normal laboratory apparatus, and

- hydrogen peroxide solution

- three powders, **P**, **Q** and **R**.

1.1   Plan an experiment to investigate which of the powders **P**, **Q** and **R** are catalysts for the decomposition of hydrogen peroxide.

- Name the apparatus used.

- Comment on how to carry out a fair test.

- Explain how you would analyse the results to see which of **P**, **Q** and **R** are catalysts.

.... (6 marks)

① Now give your answer. ✎ Then check your work against the checklist.

....
....
....
....
....
....

Checklist	
Have you included the following?	✓
measured out the hydrogen peroxide?	
named the apparatus?	
measured the volume of oxygen?	
carried out the experiment with no catalyst?	
repeated the experiment with **P/Q/R**?	
said how you would keep everything the same each time?	
said how the results would be different with a catalyst?	

....
....
....
....
....
....
....
....

Need more practice?

In the exam, extended response questions could occur as:

- simple standalone questions
- part of a question on, for example, chemical reaction
- part of a question about a practical experiment.

Have a go at this exam-style question. Write ✏ your answer on a separate piece of paper.

Exam-style question

1 In the United Kingdom, ground water, lakes and rivers are used as sources of water for making potable water.

In Singapore, a desalination plant has been built that uses sea water as a source for making potable water. The diagram shows a simple desalination process.

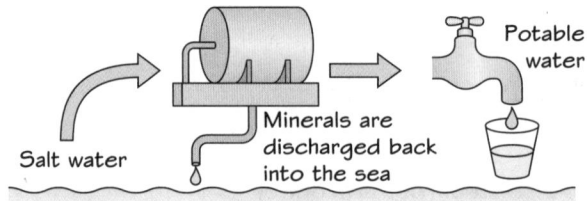

1.1 Explain why different methods of making potable water are used in different countries.

- Give basic details about making potable water from fresh water and from sea water.

- Explain why the United Kingdom does not make potable water from sea water.

- Explain why other countries have to make potable water from sea water. **(6 marks)**

Boost your grade

To improve your grade, make sure you practise extended response questions. You can find examples in the sample papers on the AQA website. You can check whether you have included all of the points mentioned by looking at the mark scheme. If you have not included all of the points, then try rewriting the answer to include them.

How confident do you feel about each of these **skills?** Colour in ✏ the bars.

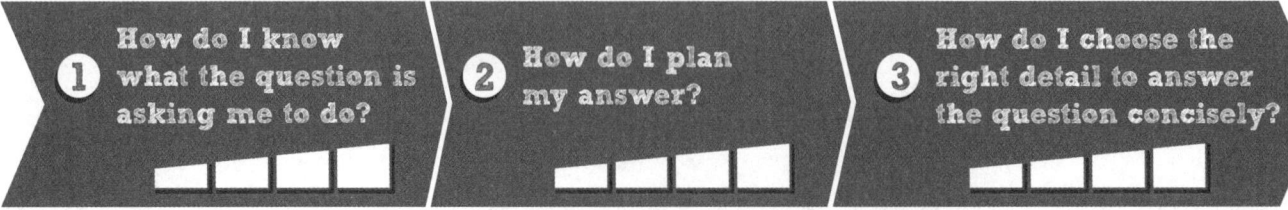

① Explaining motion using forces

This unit will help you to explain the motion of objects using the idea of forces.

In the exam you will be asked to tackle questions such as the one below.

Exam-style question

1 A skydiver jumps out of a plane and falls towards the ground.

The skydiver reaches a maximum speed.

1.1 Explain why.

.... **(4 marks)**

You will already have done some work on forces. Before starting the **skills boosts**, rate your confidence with these questions. Colour in ✐ the bars.

① **How do I identify forces on an object?**

② **How do I describe the effect of a resultant force?**

③ **How do I apply my knowledge of forces to an unfamiliar situation?**

Physics

A force is a push or pull on an object. A force can change the movement of the object it is pushing or pulling.

Remember Force is measured in newtons.

This diagram is called a **free-body diagram** (or sometimes just a force diagram).

It shows the forces acting on a skydiver.

① Label ✏ the part of the diagram that represents the skydiver.

D

W

② The downwards arrow represents the force of gravity on the skydiver.

The upwards arrow represents air resistance.

ⓐ Which term is best to describe the force of gravity on an object?
Circle Ⓐ it.

| gravity | downthrust | weight | mass |

ⓑ The downwards force is bigger.

Explain how you know this from the diagram. ✏

Think about the size of the arrows.

....

....

③ A student has written this description of the skydiver's motion.

> *Force W is bigger than force D so the skydiver is speeding up.*

ⓐ Complete ✏ the sentence below by choosing the right word(s) from the box.

| the same | opposite |

The direction of the bigger force and the direction of the skydiver's movement are

....

ⓑ Suggest a word the student could have used instead of 'speeding up' to improve their answer. ✏

....

④ Which sentence describes the forces on the skydiver when they have reached their maximum speed?

Tick ✓ **one** box.

Force W must be bigger to keep them going down. ☐

Force W and force D must be equal because the speed is not changing. ☐

Force D must be bigger because there will be a lot of air resistance at top speed. ☐

1 How do I identify forces on an object?

Before you can answer a question about forces you need to be able to identify the forces acting on an object.

A force is a push or a pull. Objects usually only push or pull one another when they are **touching**.

Look at this diagram of a horse pulling a boat using a rope.

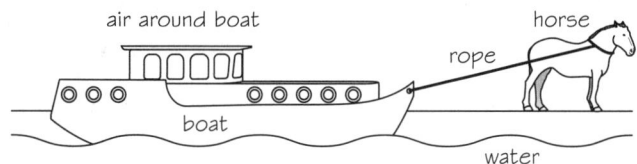

air around boat   horse   rope   boat   water

① **a** Note down ✐ in the table **three** objects (or materials) that are touching the boat.

b For each object or material you list, decide which direction they are pushing or pulling the boat. Circle Ⓐ your choice.

Object / material	Direction of push / pull
	Left / right / up / down
	Left / right / up / down
	Left / right / up / down

② Some forces can act on an object *without* touching it: they can 'act at a distance'.

a Circle Ⓐ the forces that act at a distance.

> friction   gravity   air resistance   magnetism   tension

b Which of these 'act at a distance' forces is also pulling on the boat? ✐

Think about the boat's weight.

③ Draw ✐ a diagram showing the direction of the pushes and pulls from air, water, rope and gravity.

④ The diagram shows an experiment where a paperclip is held above a table by a magnet.

List ✐ the pushes and pulls on the paperclip.

Think about what is touching the paperclip and also any 'act at a distance' forces.

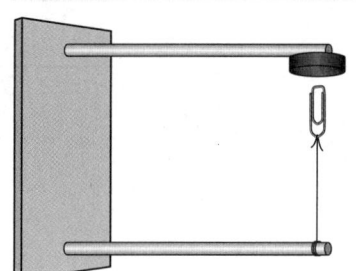

Physics

2 How do I describe the effect of a resultant force?

To work out the effect of the forces on the movement of an object you need to understand about resultant force.

On the previous page you identified the forces acting on a boat moving to the right. Here is a free-body diagram for the boat. Ignore the upward and downward forces.

Drag ←——•——→ Tension (rope)
500 N   500 N

When the forces on an object are **balanced** its speed does not change.

In this diagram the drag force and tension force are **balanced**.

1 What will happen to the speed of the boat? Circle Ⓐ the correct answer.

> increases   decreases   stays the same

2 The horse pulls harder so the tension force is now 600 N. The drag is still the same.

What will happen to the speed of the boat now? Circle Ⓐ the correct answer.

> increases   decreases   stays the same

3 When there are two or more forces on an object like this, they can be combined into a single force called the **resultant force**. Calculate ✎ the resultant force:

Add the forces if they are in the same direction; subtract one from the other if they are in opposite directions.

Resultant force =

....

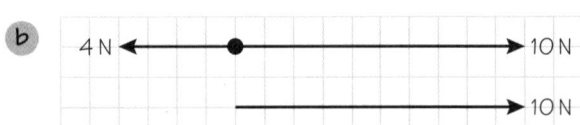

Resultant force =

....

4 **a** Show the resultant force in each of these free-body diagrams for the boat.

In each case, draw ✎ an arrow and label it with the size of the resultant force.

A 500 N ←——•——→ 600 N   B 200 N ←•——→ 500 N   C 900 N ←————•——→ 500 N

b The boat is moving to the right.

i Which diagram(s) could be correct if the boat is speeding up? Circle Ⓐ the correct answer(s).

A   B   C

ii Which diagram(s) could be correct if the boat is slowing down? Circle Ⓐ the correct answer(s).

A   B   C

5 When the resultant force is zero, what will happen to the speed of the boat? Circle Ⓐ the correct answer.

If the resultant force is zero the forces are balanced.

> increases   decreases   stays the same

Remember When there is a resultant force on an object, it accelerates. The unit of acceleration is m/s².

③ How do I apply my knowledge of forces to an unfamiliar situation?

To explain how the motion of an object changes because of the forces on it you need to identify the forces and work out the resultant force.

Air resistance is one kind of drag force. When an object moves through a material like air or water it crashes into the particles of that material. They push against the moving object.

① Draw ✎ a free-body diagram for a car travelling along to the right and pushing through the air.

> A free-body diagram consists of a dot to represent the object and arrows to represent the forces. Air resistance pushes against the direction the object is moving.

② When the car speeds up it hits more air particles each second, and it hits them harder.

What will happen to the air resistance force? Circle Ⓐ the correct answer.

> increases   decreases   stays the same

③ **a** Look at this free-body diagram for the car moving to the right.

What is the direction of the resultant force? ✎ ....

b A little while later, the forces on the car have changed:

i How does the resultant force in the diagram in **b** compare to the resultant force in the diagram in **a**? Circle Ⓐ the correct answer.

> smaller   the same   bigger

ii How is the car's speed changing in the diagram in **b**? Circle Ⓐ the correct answer.

> staying the same   still increasing   decreasing

iii What is the car doing in the diagram in **b**? Circle Ⓐ the correct answer.

> still accelerating   stationary   staying at a steady speed

> When the speed is changing an object is accelerating.

④ If the car goes fast enough, the air resistance will be as big as the forward force from the engine.

a Draw ✎ a free-body diagram to show this.

b What is the resultant force when the air resistance is the same size as the forward force? ✎

....

c What will happen to the speed of the car now? ✎

> When the forces are balanced the speed does not change.

....

Physics

Sample response

To answer a question using your ideas about forces, you need to:

- decide what forces are acting
- describe the resultant force and whether it is changing
- explain how resultant force causes acceleration which changes the motion.

Now look again at the exam-style task.

Exam-style question

1   A skydiver jumps out of a plane and falls towards the ground.

   The skydiver reaches a maximum speed.

   1.1 Explain why.

   .... **(4 marks)**

Look at these students' answers (**A** and **B**).

A

When he first jumps out, gravity pulls him down and he speeds up but then air resistance starts to slow him down until the forces are balanced and he reaches his terminal velocity.

B

When he first jumps out, his weight is much bigger than the air resistance, the resultant force is large and he has a large acceleration; as he speeds up, the air resistance increases until it balances the weight.

(1) Underline Ⓐ the names of two forces in answer **A**.

(2) Which terms are used in answer **B** but not in answer **A**? ✎

....

....

(3) Which term is used in answer **A** but not in answer **B**? ✎

....

(4) Explain which answer you think is better. ✎

....

....

....

....

....

Your turn!

It is now time to use what you have learned to answer the exam-style question below. Remember to read the question thoroughly, looking for clues. Make good use of your knowledge from other areas of physics.

Read the exam-style question and answer it using the guided steps below.

Exam-style question

1   A skydiver jumps out of a plane and falls towards the ground.

The skydiver reaches a maximum speed.

Explain means that you must say how or why something happens.

1.1   Explain why.

.... **(4 marks)**

(1) Which forces act on the skydiver? ✎

....

(2) Draw ✎ a free-body diagram for the skydiver.

(3) What happens to the size of each force as the skydiver speeds up? Circle Ⓐ the correct answers.

a   Upward force   decreases   stays the same   increases

b   Downward force   decreases   stays the same   increases

(4) What happens to the resultant force as the skydiver speeds up? Tick ✓ the correct answer.

decreases ☐   stays the same ☐   increases ☐

(5) What is the resultant force when the skydiver is at maximum speed? ✎

....

(6) What happens to the acceleration as the skydiver speeds up? Circle Ⓐ the correct answer.

decreases   stays the same   increases

(7) What is the acceleration when the skydiver is at her maximum speed? ✎

....

(8) On paper, use your answers to write ✎ a complete response to the exam-style question.

Physics Unit 1 Explaining motion using forces   **119**

Need more practice?

Exam questions often ask about different parts of a topic. Questions about forces could occur as:

- simple standalone questions

- part of a question on, for example, momentum

- part of a question about a practical test.

Have a go at this exam-style question.

Exam-style question

1   The diagram shows a steel ball falling through oil.

The ball accelerates then reaches a maximum speed.

   1.1 Explain why.

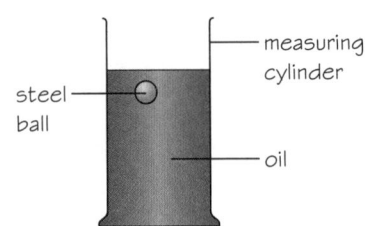

....

....

....

....

....

....

....

....

.... (4 marks)

Boost your grade

You can now identify and describe forces on an object and describe how the object's movement may change. To boost your grade, as well as describing this acceleration, you will also need to be able to calculate it using the formula $F = ma$.

How confident do you feel about each of these **skills?** Colour in the bars.

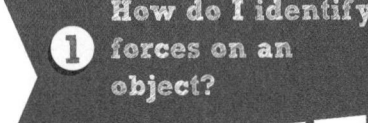

1 How do I identify forces on an object?

2 How do I describe the effect of a resultant force?

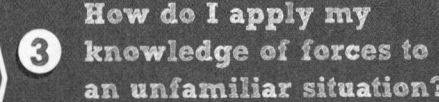

3 How do I apply my knowledge of forces to an unfamiliar situation?

② Hazards of radioactive materials

This unit will prepare you to explain the risks of using radioactive materials.

In the exam, you will be asked to tackle questions such as the one below.

Exam-style question

1 Technetium-99 is a radioactive isotope used to study blood flow in hospital patients.

Technetium-99 emits gamma radiation and has a half-life of 6 hours.

Technetium-99 is injected into the bloodstream and the gamma radiation is detected by a special camera outside the body.

A sample of technetium-99 has activity 800 MBq.

1.1 Calculate the activity of the sample after 24 hours.

.... **(2 marks)**

1.2 Explain how to minimise the risks to doctors and patients associated with using technetium-99.

.... **(4 marks)**

You will already have done some work on radioactive materials. Before starting the **skills boosts**, rate your confidence with these questions. Colour in 🖉 the bars.

1 How do I distinguish between radiation and its source?

2 How do I calculate half-life?

3 How do I describe the risks of using radioactive materials?

Radioactive materials are made from atoms whose nuclei emit radiation. Here is a reminder of some basic ideas about atoms and radiation.

(1) The nucleus of atoms contains two kinds of particle. Circle (A) the correct ones.

| electrons | neutrons | photons | protons |

A given element always has the same number of protons in its nucleus: for example, all carbon atoms have 12 protons in their nucleus. Atoms of a given element, like carbon, come in different forms called **isotopes**.

(2) What are isotopes?

> Think about the numbers of protons and neutrons that make up the nucleus.

Some isotopes are unstable. They become stable by emitting radiation.

(3) There are three main kinds of radiation from the nucleus. Draw lines to match each type of radiation with its description.

Radiation type		Description
alpha		electromagnetic radiation
beta		a high-speed electron
gamma		two protons and two neutrons, the same as a helium nucleus

(4) Each type of radiation has two important properties. Match the property to the description.

Property		Description
penetrating		can easily remove electrons from atoms they collide with
ionising		can easily pass through materials

You can think of these properties as being **opposite**. The most penetrating radiation (gamma) is the least ionising (and vice versa).

(5) Write a number (1, 2 or 3) next to each kind of radiation to show which is most (1) and least (3) ionising.

alpha ☐   beta ☐   gamma ☐

Ionisation in the cells of living things can cause damage that could increase the risk of cancer developing in the future. The amount of radiation absorbed by the body is called the **dose**: the higher the dose, the greater the risk.

(6) Write two ways that you could reduce your dose if you found yourself in a room with a radioactive source emitting gamma rays.

> You need to make sure less gamma radiation reaches your body.

1 How do I distinguish between radiation and its source?

Students often confuse a source of radiation with the radiation itself. This skills boost will help you to recognise the difference.

Look at this diagram of an unstable nucleus emitting radiation.

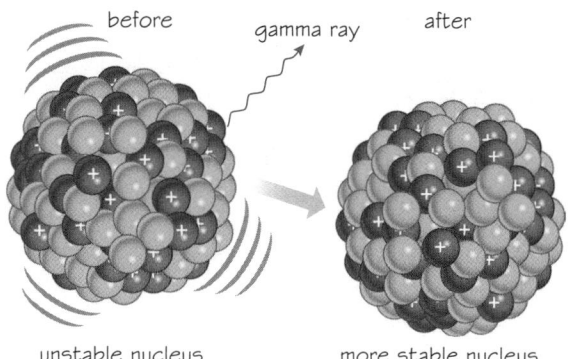

before   gamma ray   after

unstable nucleus   more stable nucleus

Any object or material that contains unstable nuclei is a **source** of radiation.

(1) Complete ⊘ the sentence using words from the box below.

The .... travels outwards from the ....

into the ....

| air | source | radioactive material | radiation | surroundings |

You will already have studied some situations involving radiation.

(2) Which of these are sources of radiation? Tick ✓ the correct ones.

rocks ☐   nuclear power stations ☐   smoke alarms ☐   Geiger counters ☐

When an unwanted source of radiation is on or in an object, we say the object is **contaminated**. For example, if some radioactive dust got onto your clothes then the clothes would be contaminated.

(3) Contaminated objects emit radiation because they contain radioactive sources.

Which of these are examples of contamination? Tick ✓ the correct ones.

| A worker at a factory accidentally breathing in radioactive dust. ☐ | A jogger running past a nuclear power station. ☐ | A hospital patient injected with a radioactive material. ☐ | The doctor who injects the patient. ☐ |

(4) A student is explaining how to work safely with a radioactive material.

> *You should wear a mask when working with radioactive materials so you don't get the radiation in you.*

The student has only been awarded one out of two marks available. Rewrite ⊘ the highlighted part of their answer to get the second mark.

....

Physics

2 How do I calculate half-life?

The risk from irradiation when you are near a radioactive source depends on its **half-life**. This page will help you understand what this means and how to find the half-life from data or from a graph.

(1) The activity of a source is 800 Bq. After 6 hours it is 200 Bq. Find the half-life. Complete 🖉 the student's workings.

> The **time** it takes for **half** of the unstable nuclei in a source to become stable is called the **half-life**.

> Count how many times you need to halve to get from 800 to 200.
>
> $800 \rightarrow \boxed{} \rightarrow 200$ There are two arrows so $\boxed{}$ half-lives.
>
> $\div 2 \left(\dfrac{2 \text{ half-lives are} \quad \boxed{} \text{ hours}}{\boxed{} \text{ half-life is} \quad 3 \text{ hours}} \right) \div \boxed{}$
>
> Answer: $\boxed{}$ hours

(2) Calculate 🖉 the half-life if the activity of a source is:

(a) 1200 Bq and after 6 hours it is 300 Bq

....

(b) 24 kBq and after 14.3 days it is 1500 Bq

....

....

> **Remember** k stands for thousand and M for million.

(c) 2 MBq and six months ago it was 4 MBq.

....

....

> In this question the numbers are the other way around.

You can also work out the future activity using the half-life. Look at this question.

(3) The activity of a source is 640 kBq. The half-life is 3 days. Calculate 🖉 the activity after 12 days by completing the student's workings.

	$\boxed{}$ days	+	3 days	+	3 days	+	3 days	= $\boxed{}$ days
640	$\rightarrow$	320	$\rightarrow$	$\boxed{}$	$\rightarrow$	$\boxed{}$	$\rightarrow$	40

Answer: $\boxed{}$ kBq

(4) The activity of a source is 1200 Bq. The half-life is 4 days. Find the activity after 12 days. 🖉

		+		+		= 12 days
	$\rightarrow$		$\rightarrow$		$\rightarrow$	

Answer .... Bq

③ How do I describe the risks of using radioactive materials?

This page shows you how to build on the ideas from the previous pages to answer questions about the risks of using radioactive materials.

① Tick ✓ which of these would reduce the risk from **irradiation**.

a. Moving away from the source ☐

b. Putting a shielding material between you and the source ☐

c. Wearing protective clothing, e.g. gloves, mask ☐

d. Using tongs to handle the source ☐

e. Only staying close to the source for a short time ☐

> Look back to page 122 to check why radiation is a risk to humans.

② If you come into contact with a radioactive source there is also a risk of **contamination**. Put one tick ✓ in each row to show which kind of radioactive source is most dangerous when inside or outside the body.

	Source emitting alpha radiation	Source emitting beta radiation	Source emitting gamma radiation
Source on your clothes	☐	☐	☐
Source inside the body	☐	☐	☐

> Alpha radiation cannot pass through skin; beta radiation can pass through skin but not clothes; gamma radiation is very penetrating.

Look at this exam-style question.

Exam-style question

1 Americium-241 is a radioactive isotope that emits alpha particles. It has a half-life of 432 years. Americium-241 is used in smoke alarms. It is put into the alarms when they are made in a factory.

Workers in the factory wear gloves and a mask.

1.1 Explain how the gloves and mask protect workers in the factory from contamination.

(2 marks)

③ Tick ✓ which of the following are examples of contamination.

a. Breathing-in the americium dust ☐   b. Standing close to the americium ☐

c. Getting americium on your clothes ☐

④ Now answer the exam-style question. ✎

....

....

....

Physics

Sample response

To answer a question about the risks of working with radioactive materials, you need to:

- decide whether irradiation or contamination is involved
- identify the type of radiation and how ionising it is
- work out or use the half-life value
- explain how this might cause ionisation in the cells of a living thing.

Now look again at the exam-style task on page 121.

(1) Look at this student's answer to part **1.1**.

> $6 + 6 + 6 + 6 = 24$ hours so 24 hours is 4 half lives
>
> Halving four times: $800 \rightarrow 400 \rightarrow 200 \rightarrow 100 \, MBq$

The number of arrows should be the number of half-lives.

a They earn one mark because one line of their answer is correct. Tick ✓ the correct line.

b Explain ✐ what mistake they have made. ....

(2) Look at this student's answer to part **1.2**.

> The doctor will be irradiated because gamma radiation is ionising and can travel a long way.
> She should wear gloves when injecting the technetium-99.
> It is a risk for the patient because they will get radiation inside them.
> It is a low risk because the half-life is 6 hours so the radiation will be gone after 12 hours.

a Circle Ⓐ two keywords used correctly in the first sentence.

b Complete ✐ this sentence:

Gloves would not offer much protection from irradiation because gamma radiation is very

....

c Explain ✐ why the third line in the student's answer is badly worded.

....

....

Think about the difference between the keywords **source** and **radiation**.

d Explain ✐ why it is wrong to say the radiation is gone in 12 hours.

....

After two half-lives the radiation level is not zero.

(3) **a** Here is another student answer. Underline Ⓐ the four points they make.

> The doctor is at risk because she might use technetium-99 every day and might get it on her clothes. The patient is at risk because the radiation will be in them for six hours and radiation causes cancer so they might die from it.

b How many marks would you give this answer? ✐ The question asked how to *minimise* the risks.

....

Your turn!

It is now time to use what you have learned to answer the exam-style question.

Read the exam-style question and answer it using the guided steps below.

Exam-style question

1 Technetium-99 is a radioactive isotope used to study blood flow in hospital patients.

Technetium-99 emits gamma radiation and has a half-life of 6 hours.

Technetium-99 is injected into the bloodstream and the gamma radiation is detected by a special camera outside the body.

A sample of technetium-99 has activity 800 MBq.

1.1 Calculate the activity of the sample after 24 hours.

.... (2 marks)

1.2 Explain how to minimise the risks to doctors and patients associated with using technetium-99.

.... (4 marks)

1.1 i Circle Ⓐ the initial activity and half-life given in the question.

ii Complete 🖉 the boxes to find the answer.

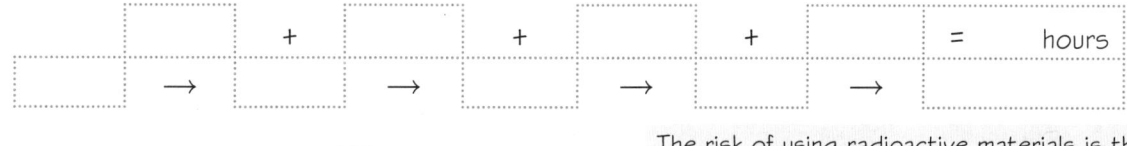

.... MBq

The risk of using radioactive materials is that radiation is ionising and can damage cells.

1.2 i Circle Ⓐ who will be at risk due to contamination.   doctor   patient

ii How can their radiation dose be made smaller? ....

iii How could the doctor become irradiated? 🖉 ....

iv How could they reduce their dose? 🖉 ....

v How could the doctor become contaminated? 🖉 ....

vi How could they reduce the risk of becoming contaminated? 🖉 ....

....

vii Now use your answers to write 🖉 a complete response to part 1.2.

....

....

....

....

....

Need more practice?

In the exam, questions about hazards of radioactive materials could occur as:

- simple standalone questions
- part of a question on hazards of radioactive material
- part of a question about a practical test.

Have a go at this exam-style question.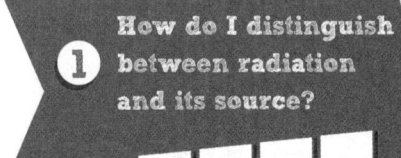

Exam-style question

1   Old nuclear power stations contain radioactive materials.

    1.1   Explain why robots rather than people are used to help demolish old nuclear power stations.

       (4 marks)

Boost your grade

You can now describe and explain the risks of using radioactive materials and how to minimise those risks. You should also be able to write equations to show what happens to the unstable nuclei.

How confident do you feel about each of these **skills?** Colour in 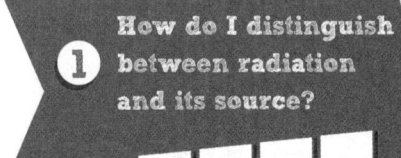 the bars.

1 How do I distinguish between radiation and its source?

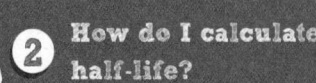
2 How do I calculate half-life?

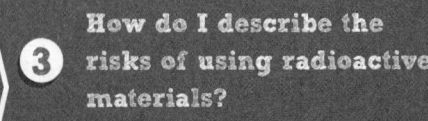
3 How do I describe the risks of using radioactive materials?

③ Heating and kinetic theory

This unit will help you to explain the properties of materials using the kinetic theory model in any given situation.

In the exam you will be asked to tackle questions such as the one below.

Exam-style question

1   A fixed mass of gas is contained within a rigid metal container of fixed volume.

   1.1   Explain how the pressure exerted by the gas changes when the gas is heated.

   .... **(3 marks)**

You will already have done some work on the kinetic theory model. Before starting the **skills boosts**, rate your confidence with these questions. Colour in 🖉 the bars.

① How do I describe the kinetic theory model?

② How do I describe what happens during heating?

③ How do I explain how a gas exerts a pressure on its container?

The particles in solids, liquids and gases are arranged and move differently.

(1) Label ✐ the diagrams using the words in the word bank.

| solid | liquid | gas |

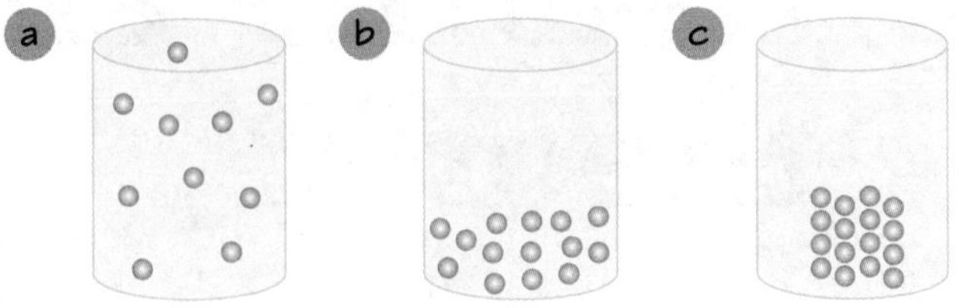

a   b   c

....

(2) How are the particles moving in each state? Match ✐ the boxes below.

Solid			Rapid, random motion
Liquid			Vibrate around fixed positions
Gas			Move past one another

(3) Solid, liquid and gas are called the three **states of matter**. Match ✐ the boxes below to show what the changes of states of matter are called.

Melting			Turning straight from solid into gas
Boiling			Turning from liquid into gas
Sublimation			Turning from solid into liquid

(4) The opposite of sublimation is **desublimation**. Complete ✐ the words for the opposites of boiling and melting.

fr............................   con............................

(5) Here are two pictures of the particles of a gas.

In **Figure 1**, the particles are moving faster on average than in **Figure 2**.

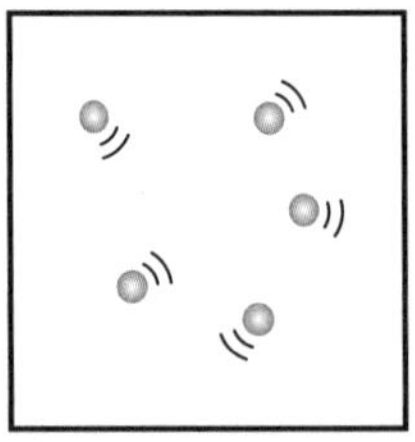

Figure 1 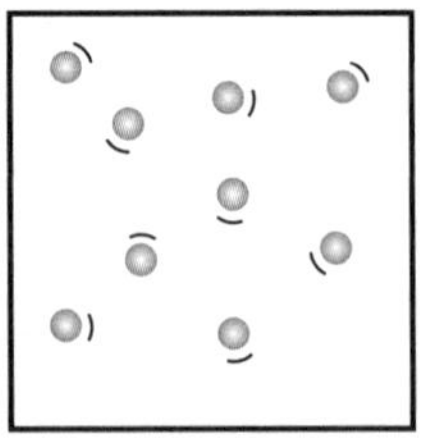   Figure 2

(a) Complete ✐ the sentence using words from the box.

The particles in **Figure 1** have a higher average ....

| kinetic energy | thermal energy | internal energy | temperature |

(b) Complete ✐ this sentence using words from the box.

The **gas** in **Figure 1** has a lower ....

| weight | density | volume | average kinetic energy |

1 **How do I describe the kinetic theory model?**

Students often mix up the words used to describe particles and the words used to describe a substance. This page will help you practise choosing the right words.

Look again at this picture of the particles in a gas.

The particles are far apart and moving around quickly.

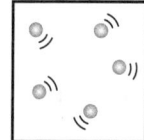

① Read these sentences about **gas particles**. The words in bold should only be used to describe particles. The words in the box should only be used to describe the gas. Choose the correct word from the box to complete 🖉 each sentence.

volume	density	temperature	expands

a The particles have a lot of **kinetic energy**; the gas has a high ....

b The particles **move further apart**; the gas ....

c The particles **are far apart**; the gas has a low ....

d The particles are in a closed container; the gas has a fixed ....

The properties of a substance can be explained using ideas about the particles.

② Draw 🖉 lines to match the property with the explanation.

Property

| A gas expands to fill its container because |
| A solid cannot flow because |
| A liquid has a high density because |

Explanation

| particles vibrate around fixed positions. |
| particles are packed closely together. |
| there are no forces between the particles. |

③ A student has written the following description of particle motion in a gas.

| A gas has <u>a lot of kinetic energy</u> because the particles are moving fast. |

The first exercise on this page shows which words to use about a gas.

a Rewrite 🖉 the underlined part of the sentence choosing better words to describe the gas.

....

Here is another part of the student's answer.

| If the metal bar is heated the particles expand and eventually it melts. |

A better answer would be:

| If the metal bar is heated it expands because the particles move apart. Eventually the bar melts. |

b Explain 🖉 why the second answer is better.

....

....

Physics

Physics Unit 3 Heating and kinetic theory 131

2 How do I describe what happens during heating?

This page will help you to describe how the arrangement and movement of particles change when the substance is heated.

The total energy of the particles of a substance is called its **internal** energy. The internal energy is made up of:

- **kinetic** energy of the particles, because they are moving
- **potential** energy of the particles, if they are attracted to each other.

(1) Complete ✏ the sentence using the words in bold above.

The .... energy of a substance is equal to the .... energy

of its particles plus the .... energy of its particles.

Two things can happen when a substance is heated:

- The particles gain **kinetic energy** (they move more); the substance's **temperature** increases.
- The particles gain **potential energy** (they move apart); the substance changes state (e.g. melts or boils).

(2) Complete ✏ the sentences below using the words in bold above.

When a substance is melting, its particles are gaining .... and its

.... does not change. When the temperature of a substance is increasing,

its particles are gaining ....

(3) Look at the graph of the temperature of a substance being heated.

a Write ✏ MELT on the graph where the substance is melting.

The temperature doesn't change when the state is changing.

b Write ✏ KE INCREASE on the graph where the particles' kinetic energy is increasing.

When the particles' kinetic energy (KE) increases the substance's temperature increases.

(4) Circle Ⓐ the correct word to complete each sentence.

a When a substance is heated its internal energy **increases** / **decreases** / **stays the same**.

b When a substance changes state the temperature **increases** / **decreases** / **stays the same**.

c Temperature depends only on the particles' **kinetic energy** / **potential energy** / **internal energy**.

Now use these ideas to answer the following questions.

(5) Water at 10°C is heated. Explain ✏ using ideas about particles why the temperature of the water increases.

....

(6) Ice at 0°C is heated. It turns into water at 0°C. Explain ✏ using ideas about particles why the temperature stays the same.

....

....

3 How do I explain how a gas exerts a pressure on its container?

This skills boost shows you how to combine the ideas from the previous two skills boosts to answer questions about the pressure of a gas.

(1) How are the particles in a gas moving? ✎ — Look back to pages 130 and 131.

....

When the particles in a gas collide with the inside of the walls of a container and bounce back, the particles push the walls outwards. This is the gas pressure.

(2) Circle Ⓐ the correct word to complete each sentence.

a Use **pressure** when describing the **gas** / **particles**.

b Use **collisions** when describing the **gas** / **particles**.

(3) Complete ✎ this sentence. Include the keywords **colliding** and **particles**.

The pressure of a gas on its container is caused by ....

(4) Which factors do you think would affect the pressure of the gas? Tick ✓ the correct answer(s).

speed of particles ☐   mass of particles ☐

number of particles ☐   volume of container ☐

(5) Which of the factors in (4) will change if you heat the gas? ✎ ....

(6) Which of the factors in (4) will change if you pump more gas into the same container? ✎

....

(7) Use your answers to plan a response to this exam-style question then write your answer on paper. ✎

Exam-style question

1 A student has a bicycle with a flat tyre.

The student uses a pump to increase the pressure.

1.1 Explain how pumping more air into the tyre increases the pressure.

....

....

....

....   (3 marks)

Sample response

Now look again at this exam-style task and some student responses to the question.

Exam-style question

1   A fixed mass of gas is contained within a rigid metal container of fixed volume.

   1.1   Explain how the pressure exerted by the gas changes when the gas is heated.

   .... (3 marks)

(1) Here is one student's response.

A	*Hotter particles move faster so they hit the walls harder.*

This student may not earn any marks because he has used everyday words instead of technical terms. Annotate 🖉 answer A to improve it, suggesting technical terms to replace the words: 'Hotter particles' and 'hit'.

(2) Here is another student's answer.

B	*The pressure increases when the gas is heated because the particles collide with the walls more often.*

This student would earn 2 marks. Underline Ⓐ the two parts of their answer that you think are worth a mark.

(3) Answer **B** uses technical language but it does not explain why the particles collide more often when the gas is heated. Complete 🖉 this sentence using what you have learned from the previous activities.

Heating the gas makes the particles collide with the walls more often because ....

....

....

                                           What happens to the particles' average speed?

(4) Complete 🖉 this sentence.

If more particles with the same average energy are added to the box, the pressure increases

because ....

....

Your turn!

It is now time to use what you have learned to answer the exam-style question. Remember to read the question thoroughly, looking for clues. Make good use of your knowledge from other areas of physics.

Read the exam-style question and answer it using the guided steps below.

Exam-style question

1   A fixed mass of gas is contained within a rigid metal container of fixed volume.

   1.1   Explain how the pressure exerted by the gas changes when the gas is heated.

   .... (3 marks)

(1) What causes gas pressure? ✎ ....

....

(2) When the gas is heated, what happens to the particles? ✎

....

....

(3) What happens to the gas pressure when the gas is heated? Circle Ⓐ the correct answer

   ┌───┐
   │   increases   stays the same   decreases   │
   └───┘

(4) Explain ✎ your answer to (3) using your answer to (2).

....

....

(5) Now use your answers to write ✎ a complete response to the exam-style question.

....

....

....

....

....

....

....

(6) Why was it important that the question said the mass of gas and volume of the container were fixed? ✎

....

....

Physics

Need more practice?

Exam questions often ask about different parts of a topic. Questions about kinetic theory could occur as:

- simple standalone questions
- part of a question on how energy transfers on heating
- part of a question about a practical test.

In this section you are going to review your skills and have an opportunity to practise applying them to a new situation.

Have a go at these exam-style questions. If you need more space to write your answer, continue on paper.

Exam-style questions

1   A can containing a gas carries a warning: 'Do not dispose of in fire.'

The gas does not burn.

1.1   Explain why it is dangerous for the can of gas to get very hot.

....

....

....

.... (3 marks)

2   A student hangs some washing out to dry.

Her dad tells her to spread out the clothes on the washing line.

2.1   Explain why this will help the clothes to dry faster.

....

....

....

.... (3 marks)

Boost your grade

You can now use the kinetic theory model to describe and explain the properties of solids, liquids and gases. You should also be able to evaluate ways to increase evaporation.

How confident do you feel about each of these **skills?** Colour in the bars.

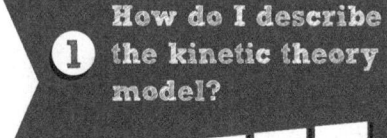

① How do I describe the kinetic theory model?

② How do I describe what happens during heating?

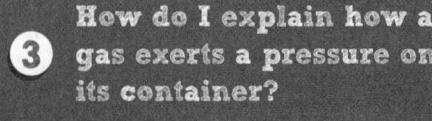

③ How do I explain how a gas exerts a pressure on its container?

④ Electrical resistance and safety

This unit will help you to explain electrical resistance and how to use electricity at home.

In the exam you will be asked to tackle questions such as the one below.

1 The plug for a washing machine contains a fuse. The fuse is a safety device.

   1.1 Explain how the fuse makes the washing machine safe if there is a fault.

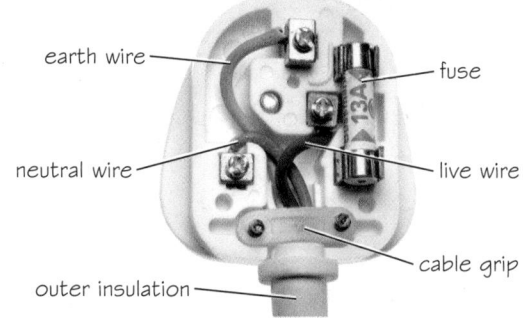

earth wire ⎯ fuse

neutral wire ⎯ live wire

cable grip

outer insulation ⎯

.... (3 marks)

You will already have done some work on this. Before starting the **skills boosts**, rate your confidence with these questions. Colour in 🖊 the bars.

1 How do I distinguish current from potential difference (voltage)?

2 How do I explain the effect of resistance in a circuit?

3 How do I explain the safety features of the mains electricity supply?

This topic is all about how energy is transferred by electrons moving around a circuit.

The unit of charge is the coulomb; this is very often asked as part of an exam question.

(1) Draw lines ✏ to link the keyword in the centre to the correct meaning on the right and the correct unit on the left. One has been done for you.

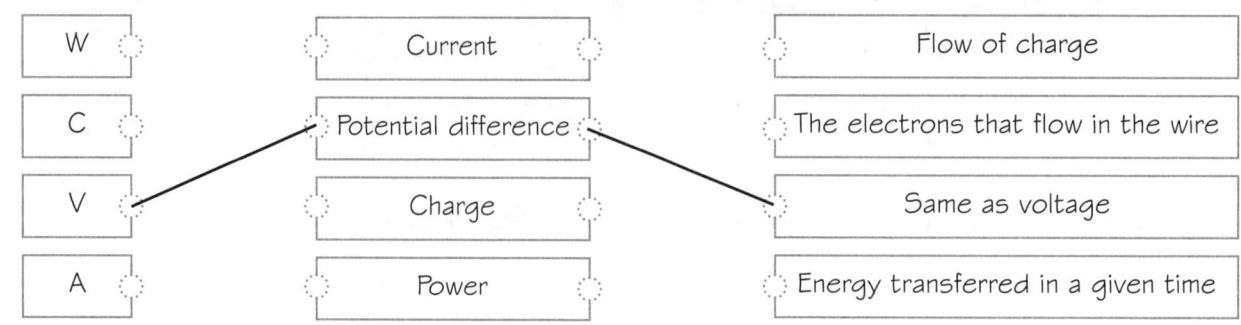

W		Current		Flow of charge
C		Potential difference		The electrons that flow in the wire
V		Charge		Same as voltage
A		Power		Energy transferred in a given time

(2) These students have used the word electricity in their answers. Annotate ✏ the answers, replacing 'electricity' with a more appropriate word.

> *Electricity can flow through the circuit easier when there's less resistance.*

Avoid using 'electricity' in your answers, use either energy, current or charge instead.

> *A normal light bulb uses more electricity each day than an LED bulb.*

It is easy to confuse symbols used in equations, such as $V = IR$, with their units (volts, V, amps, A, and ohms, Ω).

(3) **a** Underline Ⓐ the **units** in the word box for:

   i current

   ii potential difference

   iii resistance.

> amp   resistance   R   ohm
> A   I   current   Ω   potential difference
> V   voltmeter   ammeter   volt

  b Circle Ⓐ and connect the words, symbols and units that go together.

(4) Add ✏ an A and V to the circuit on the right to label the ammeter and voltmeter correctly.

(5) What is the device represented by this symbol? ✏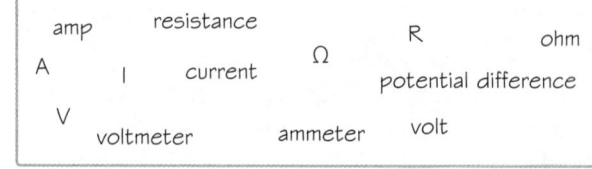

....

(6) What does the device in (5) do in the circuit? ✏

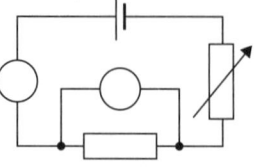

When resistance in a circuit increases, the current decreases.

....

Resistance is calculated using Ohm's law:

$$\text{resistance (in } \Omega) = \frac{\text{potential difference (in V)}}{\text{current (in A)}}$$

(7) In the device in (5), when the potential difference is 6 V the current is 0.1 A.

Calculate ✏ the resistance of the device.

1 How do I distinguish current from potential difference (voltage)?

Students often mix up the words current and potential difference (voltage) in their answers. This page will help you to understand the difference between them.

In a conductor like a copper wire, some of the electrons are free to move.

① This diagram shows the atoms in a copper wire and the free electrons that are able to move through the wire. Complete ✎ the labels using the words **copper atom** and **free electron**.

② Circle Ⓐ the correct word to complete this sentence: Electrons are **positive / neutral / negative**.

When the wire is connected to a battery, the free electrons in the wire are attracted to the positive side of the battery. This makes them move through the wire.

③ Draw an arrow ✎ on this diagram to show which way the electrons will try to move.

To understand the difference between current and potential difference (voltage) it might help to imagine each electron as a truck that can carry coal. The trucks are filled up in the battery and unload in the bulbs, resistors, etc.

• Current describes the number of trucks driving past in a given time.

• Potential difference (voltage) describes how full of coal each truck is.

④ Use this idea to help you match ✎ each keyword with its scientific meaning.

Keyword	Scientific meaning
Current	The amount of energy transferred by each charge as it passes through a device.
Potential difference (voltage)	The number of charges passing a given point in one second.

Current describes how much charge (how many electrons) pass a point in the circuit every second.

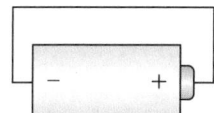

$$current\ (in\ A) = \frac{charge\ flowing\ (in\ C)}{time\ taken\ (in\ s)}$$

⑤ 3 C of charge pass through a battery in 1 minute. Calculate ✎ the current using the above formula.

Potential difference (voltage) describes how much energy is transferred to or from the electrons in part of a circuit.

$$potential\ difference\ (in\ V) = \frac{energy\ transferred\ (in\ J)}{charge\ (in\ C)}$$

⑥ 5 C of charge pass through a 9 V battery. Calculate ✎ the energy supplied by the battery to the electrons using the above formula.

Substitute into the formula and solve to find the energy.

Physics

② How do I explain the effect of resistance in a circuit?

To answer questions about circuits you will need a good understanding of resistance and how changes in resistance affect the current. This page will help you to describe current and resistance in circuits.

① Complete ✎ this sentence.

Current is a .... of ....

In a simple circuit, current is made up of negatively-charged electrons flowing through the wires.

Resistance describes how hard it is for a given battery to push current through an object.

Remember Increasing the resistance decreases the current (and vice versa)

② **a** What will happen to the current in a circuit when more devices are added? ✎ ....

....

b Explain ✎ why. ....

③ Explain ✎ how the reading on the ammeter will change when a second bulb is added to this circuit.

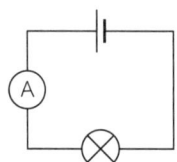

Say what happens to the reading then say why adding a bulb makes this happen.

....

....

A current in a wire also causes a heating effect which depends on the resistance. Increasing the current increases the heating effect (for a given resistance).

④ Circle Ⓐ which devices use this heating effect to do a useful job.

> toaster   wire   bulb   laptop   hairdryer

⑤ What might happen if the current becomes too high? ✎ ....

Resistance is calculated using Ohm's law:

$$\text{resistance (in } \Omega) = \frac{\text{potential difference (in V)}}{\text{current (in A)}}$$

When you use this in your exam you should specify clearly which device's resistance you are working out. Doing this will help you avoid the common mistake of using the wrong numbers in the equation.

$$\text{Resistance of bulb} = \frac{\text{PD across bulb}}{\text{current through bulb}}$$

⑥ Use the circuit shown in the diagram to complete ✎ these calculations.

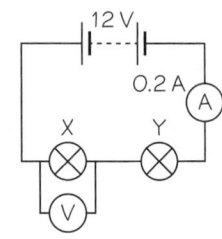

a Resistance of <u>bulb X</u>   = $\dfrac{\boxed{}}{0.2 \text{ A}}$ = ...

b <u>Total</u> resistance of circuit = $\dfrac{\boxed{}}{0.2 \text{ A}}$ = ...

3 How do I explain the safety features of the mains electricity supply?

This page will help you to understand how mains electricity differs from the circuits you build in school and how to use the ideas from the previous pages to answer questions about electrical safety.

The mains electricity supply has a higher potential difference than batteries.

1 Draw 🖉 lines to match the supplies to their typical potential differences.

| Battery | | | 230 V |
| Mains | | | 1.5 V |

2 The diagram shows the inside of a mains plug.
Complete 🖉 the gaps in the sentences below by choosing either **green/yellow**, **blue** or **brown**.

earth wire — fuse
13A
neutral wire — live wire
cable grip
outer insulation —

Use the diagram to help.

The live wire is coloured ....

The earth wire is coloured ....

The neutral wire is coloured ....

3 Complete 🖉 this sentence. The fuse is connected with the .... wire.

A fuse is a deliberate **weak link** in the circuit. It contains a thin wire that melts if it gets too hot, and this breaks the circuit.

4 Complete 🖉 the sentence using words from the box below.

If there is a fault, the current .... The thin wire in the fuse gets

very .... and .... This .... the circuit.

| increases | snaps | breaks | hot | melts | decreases | cold | blows |

If there is a fault and the appliance has metal parts on the outside that you could touch, you could be electrocuted. The earth wire works with the fuse to prevent this.

5 Two of the following are safety features. Tick ✓ which ones.

live wire ☐   earth wire ☐   neutral wire ☐   fuse ☐

Some devices use plugs and cables with only live and neutral wires. They do not need the earth wire because they have no exposed metal parts.

6 Tick ✓ the devices that should be fitted with an earth wire.   Which have exposed metal parts?

washing machine ☐   hairdryer ☐   TV ☐   iron ☐   lawnmower ☐

Physics

Sample response

To answer a question about electrical resistance and safety, you need to be able to:

- understand the difference between current and potential difference (voltage)
- explain how resistance affects the current in a circuit which causes heating
- describe how mains electricity is different and can be used safely.

Now look again at this exam-style question.

Exam-style question

1   The plug for a washing machine contains a fuse. The fuse is a safety device.

   1.1   Explain how the fuse makes the washing machine safe if there is a fault.

   ....   (3 marks)

(1) What happens to the resistance if there is a fault? ✎ ....

(2) What happens to the current? ✎ ....

(3) What happens to the wire inside the fuse? ✎ ....

Look at this student's answer.

> When there is a fault the current goes up and the fuse blows making the appliance safe.

(4) The student should have used the words **melts** and **increases**. Cross out and replace ✎ the incorrect words in the answer above.

The student has lost another mark because they haven't explained why this will make the appliance safe.

(5) Why is the appliance safe once the fuse wire has melted? ✎

Is there a complete circuit if the wire melts?

....

....

Here is another student's answer.

> The washing machine is made of metal and when the fuse blows it stops you being electrocuted.

This student has mixed-up two safety features in her answer.

(6) Which two safety features has this student mixed-up? ✎

....

(7) Draw lines ✎ to match the safety feature to the hazard.

Safety feature		Hazard
Earth wire		overheating and fire
Fuse		electrocution through the metal case

Your turn!

It is now time to use what you have learned to answer the exam-style question. Remember to thoroughly read the question and look for clues. Make good use of your knowledge from other areas of physics.

Read the exam-style question and answer it using the guided steps below.

Exam-style question

1 The plug for a washing machine contains a fuse. The fuse is a safety device.

    1.1 Explain how the fuse makes the washing machine safe if there is a fault.

    .... (3 marks)

(1) Describe 🖊 how the current changes if there is a fault.

....

....

(2) Describe 🖊 what happens inside the fuse.

....

....

(3) Describe 🖊 how this turns off the circuit.

....

....

(4) Now use your answers to write 🖊 a complete response to the exam-style question.

....

....

....

....

....

....

....

Physics

Need more practice?

Exam questions often ask about different parts of a topic. Questions about electrical resistance and mains electricity could occur as:

- simple standalone questions

- part of a question on energy transfer in the National Grid or on electromagnets

- part of a question about a practical test.

Have a go at this exam-style question.

Exam-style question

1   A shower is normally connected to the mains supply with a thick wire.

Thinner wires have higher resistance.

When the shower is working, the current in the wires is very high.

1.1   Explain why using thinner wire to connect the shower is dangerous.

....

....

....

....   (2 marks)

Boost your grade

You can now use ideas about current, potential difference and resistance to describe and explain safety features of the mains. You should also be able to describe devices that change their resistance.

How confident do you feel about each of these **skills?** Colour in 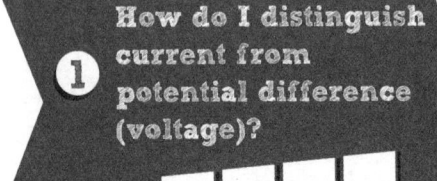 the bars.

1 How do I distinguish current from potential difference (voltage)?

2 How do I explain the effect of resistance in a circuit?

3 How do I explain the safety features of the mains electricity supply?

(5) Maths skills

This unit will help you to use maths to find the answers to physics problems.

As well as describing ideas about the universe, you need to put values to them. You can use maths to do this.

In the exam you will be asked to tackle questions such as the one below.

Exam-style question

1   A car is accelerating on a level road. The forward force on the car is 1800 N. The drag force on the car is 950 N. The car has a mass of 900 kg and the passengers inside have a total mass of 150 kg.

   1.1   Calculate the acceleration of the car. Give your answer to two significant figures.

   Give the unit.

   ....   **(3 marks)**

You will already have done some work on maths skills. Before starting the **skills boosts**, rate your confidence with these questions. Colour in the bars.

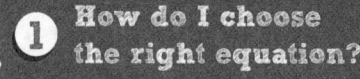

1 How do I choose the right equation?

2 How do I decide which values to use?

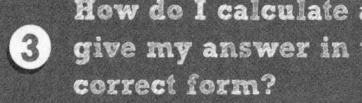

3 How do I calculate and give my answer in the correct form?

Physics

There are lots of different units for speed in everyday use. Physics uses an agreed set of units called SI units, based on units like kilograms (for mass), metres (for length) and seconds (for time).

The equation for speed is:

$$speed = \frac{distance}{time}$$

1 Tick ✓ the speed which is given with the correct SI unit.

25 cm/s ☐   mach 2 ☐   3 mph ☐   10 m/s ☐   40 km/h ☐

2 What is the SI unit for distance? ✎ ....

3 What is the SI unit for time? ✎ ....

4 A runner covers 100 m in 12 s. Calculate ✎ her speed.

Speed = $\dfrac{\boxed{}}{12}$

     = ... m/s

Substitute the values from the question to complete the calculation.

Potential difference, current and resistance are related by the equation:

$$potential\ difference = current \times resistance$$

5 What is the SI unit for current? ✎ ....

6 What is the SI unit for potential difference? ✎ ....

Potential difference is another term for voltage.

7 A current of 2 A passes through a resistor of resistance 100 Ω.

Calculate ✎ the potential difference across the resistor.

Substitute the values from the question and complete this calculation.

potential difference = ... × ....

               = ... unit ....

How do I choose the right equation?

To find the right equation you first need to identify the quantity you are being asked to calculate and the values you are given in the question.

(1) A car stops in a distance of 20 m. The work done by the braking force is 2000 N. Calculate the braking force.

a Circle Ⓐ the quantity you want to find.

b Underline Ⓐ the **names** of the two values given in the question.

c Tick ✓ which of the equations below is the correct equation to use.

Which two values do you know?

force = mass × acceleration	☐	work done = force × distance	☐
force = spring constant × extension	☐		

d Work out the braking force. 🖉

Rearrange the equation from **c** and substitute the values from the question.

Sometimes the question will ask for the **change** in a quantity. In this case you need to use the equation to work out the quantity twice: before and after.

(2) A stationary golf ball of mass 0.1 kg is hit with a club and moves off at 20 m/s. Calculate the change in momentum of the ball.

Remember
The velocity of the ball is zero when it is stationary.

Substitute the values from the question to complete 🖉 this calculation.

Before: momentum = mass × velocity = ...

After: momentum = mass × velocity = ...

Change in momentum = ... – .... = ... kg m/s

Try this one, it is a little bit more tricky.

(3) A tennis ball of mass 0.05 kg is travelling at 35 m/s. It bounces and slows down to 25 m/s. Calculate the change in kinetic energy of the ball.

Substitute the values from the question to complete 🖉 this calculation.

Before: kinetic energy = $\frac{1}{2}$ × mass × speed² = ...

= ....

After: kinetic energy = $\frac{1}{2}$ × mass × speed² = ...

= ....

Change in kinetic energy = ... – .... = ... J

2 How do I decide which numbers to use?

This page will help you to recognise which numbers to put into the equation you have chosen.

Most numbers in physics come with a unit such as J, N, kg, m or Hz. Sometimes another letter is used before the unit: this changes the size of the unit.

The kilo- (k) prefix means 1000 and the milli- (m) prefix means 1/1000th or 0.001. You also need to know the prefixes mega (M, million) and giga (G, one thousand million).

(1) Match 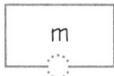 the prefixes to their values.

| m | G | M | k |

| 1000 | 1 000 000 | 1 000 000 000 | 0.001 |

Physics uses a standard set of units called SI units, based on metres, kilograms and seconds. Other units have to be converted to the standard ones.

(2) A resistor has resistance 15 kΩ. Calculate the potential difference across the resistor when the current is 10 mA.

Substitute the values from the question and complete  the calculation.

15 kΩ = ...Ω   10 mA = ...A

pd = current × resistance = ... × ....= V

(3) Calculate the weight of a 50 g mass.

Complete the calculation.

50 g = ...kg   Change to standard units before using the equation.

weight = mass in kg × 10 N/kg = ...N

(4) A wire carries a current of 5 A. Calculate the charge passing in 1 minute.

Complete  the calculation.

1 minute = ...s   Convert to standard units before using the equation.

charge = current × time = ... × .... = ...C

Sometimes you may need to calculate a quantity you need for an equation.

(5) A student of mass 60 kg is riding her bike. The bike has mass 20 kg.
Calculate the kinetic energy of the bike and rider when travelling at 10 m/s.

a First, calculate the total mass.

The total mass is .... kg + ... kg = ... kg

b Substitute the total mass into the equation.

kinetic energy = $\frac{1}{2}$ × mass × speed² = $\frac{1}{2}$ × .... × 10² = ... J

③ How do I calculate and give my answer in the correct form?

Some questions on the exam paper will ask you to give the unit and to answer with a given number of significant figures. This page will help you to calculate the answer and give the correct details.

Some calculations need you to solve an equation to find a quantity.

① Water waves with a frequency of 2 Hz have a speed of 10 m/s. Calculate the wavelength of the waves.

Complete ✎ the calculation.

wave speed = frequency × wavelength

$\div 2$ ⟨ .... = 2 × wavelength ⟩ $\div 2$

.... = wavelength   wavelength = ... m

② A skateboard is travelling at 6 m/s. The momentum of the skateboard is 30 kg m/s. Calculate the mass of the skateboard.

Work out ✎ the answer.

Remember to give the unit if it is not already there. Usually there is an extra answer line on the exam paper with the word **unit** to remind you to include it. The exception is efficiency. This has no unit (or can be written as a percentage).

③ Match ✎ the unit to the quantity.

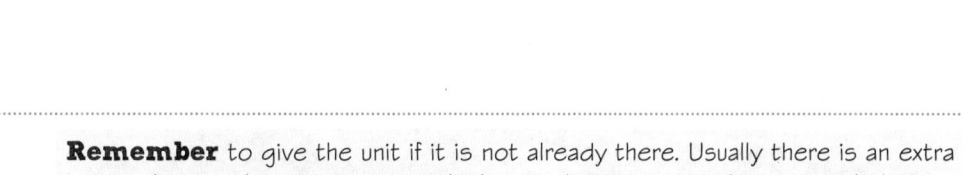

| Energy | Mass | Time | Charge | Pd | Power |

| C | W | kg | s | J | V |

④ This student answering a question on energy has made a mistake:

Energy transferred = 20 × 6 = 120 j

Capital letters and small letters may stand for different units. Learn which is correct.

Circle Ⓐ the mistake the student has made and annotate ✎ the answer to correct it.

In a question you might be asked for a certain number of significant figures.

⑤ How many significant figures have been used for each of these numbers?

Count the number of digits: only include 0s if they are after the decimal point.

　ⓐ 12 ....　　ⓑ 2.7 ....

　ⓒ 1.06 ....　　ⓓ 2.0 ....

Physics

Sample response

To answer a calculation question, you need to:

- remember or choose the correct equation to use
- identify the correct numbers from the question
- solve the equation and give the answer with the right units and significant figures.

Now look again at this exam-style task.

Exam-style question

1 A car is accelerating on a level road. The forward force on the car is 1800 N. The drag force on the car is 950 N. The car has a mass of 900 kg and the passengers inside have a total mass of 150 kg.

1.2 Calculate the acceleration of the car. Give your answer to two significant figures.

Give the unit.

.... (3 marks)

(1) Look at student A's answer. **A**

$$acceleration = \frac{change\ in\ speed}{time} = \frac{1800 - 950}{1\ minute}$$
$$= 850\ m/s$$

a Why can't the student use this equation?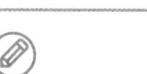

....

....

Look at the quantity you want to find and the **names** of the values given in the question.

b What other mistakes has the student made?

....

....

Look at the unit.

(2) Now look at student B's answer. **B**

$$force = mass \times acceleration$$
$$1800 = 900 \times acceleration$$
$$Acceleration = 2\ m/s^2$$
$$= 4\ m/s$$

a Explain why it is a mistake to square the answer, 2, to get 4.

....

....

b Student B has also made mistakes with the force and mass values. Describe the mistakes the student has made.

Remember the unit is just a name.

....

....

Your turn!

It is now time to use what you have learned to answer the exam-style question. Remember to thoroughly read the question and look for clues. Make good use of your knowledge from other areas of physics.

Read the exam-style question and answer it using the guided steps below.

Exam-style question

1   A car is accelerating on a level road. The forward force on the car is 1800 N. The drag force on the car is 950 N. The car has a mass of 900 kg and the passengers inside have a total mass of 150 kg.

  1.1   Calculate the acceleration of the car. Give your answer to two significant figures.

  Give the unit.

  .... (3 marks)

① List ✎ all the equations from your formula sheet that include acceleration.

② Tick ✓ which of these values are given in the question.

change in velocity ☐   time ☐   mass ☐   force ☐

③ There are two values of force in the question. What value should you use in the equation? ✎

....

The forces are in opposite directions. Forces are explained in Unit 1.

④ There are two values of mass in the question. What value should you use in the equation? ✎

....

What is the total mass of the car and passengers?

⑤ Which of these is the unit for acceleration? Circle Ⓐ the correct one.

W   m/s²   V   m/s   C   J   m

⑥ What is the acceleration of the car? ✎

Need more practice?

In the exam, questions using maths skills could occur as:

- simple standalone questions
- part of a question on any physics topic, especially calculations
- part of a question about a practical test.

Have a go at these exam-style questions.

Exam-style questions

1 A motor transfers 200 J usefully and wastes 100 J every minute.

   1.1 Calculate the power of the motor.

   Give the unit.

   ....

   ....

   ....

   ....

   ....

   Power .... Unit ....   **(4 marks)**

2 A radio station broadcasts radio waves with a wavelength of 20 m.

   The speed of light is 3.0×10^8 m/s.

   2.1 Calculate the frequency of the waves.

   ....

   ....

   ....   **(3 marks)**

Boost your grade

You can now choose the right equation and values to use for a calculation and give the unit and significant figures correctly. You should also be able to use standard form in some calculations.

How confident do you feel about each of these **skills**? Colour in 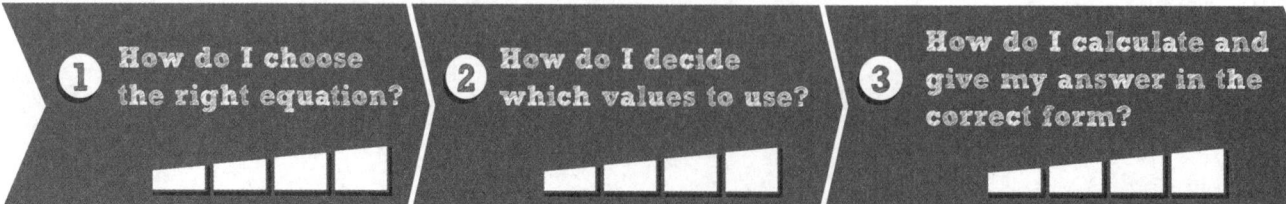 the bars.

1 How do I choose the right equation?

2 How do I decide which values to use?

3 How do I calculate and give my answer in the correct form?

⑥ Graph skills

This unit will help you to describe and use graphs correctly.

An important part of physics is describing the patterns we see in our observations about the universe. Graphs help to show those patterns.

In the exam you will be asked to tackle questions such as the one below.

1   A student is investigating the relationship between the force on a spring and its extension.

   The student's data is shown plotted as a graph in **Figure 1**.

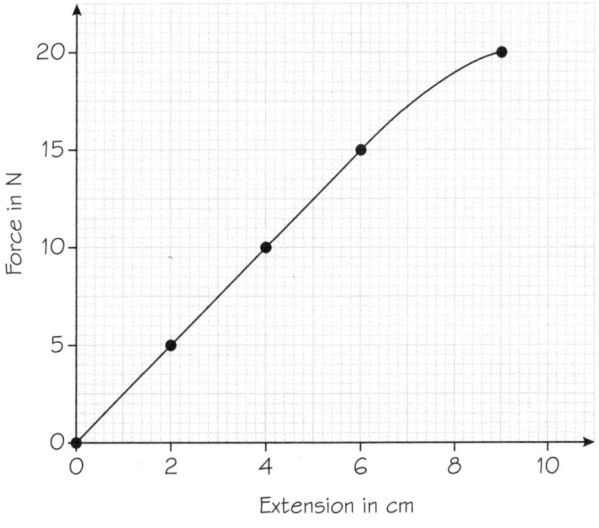

Figure 1

1.1   Describe the relationship shown by the graph.

   .... (2 marks)

1.2   Use **Figure 1** to find the extension of the spring when the force is 12 N.

   .... (2 marks)

1.3   Use **Figure 1** to find the spring constant in N/m.

   .... (2 marks)

You will already have done some work on graphs. Before starting the **skills boosts**, rate your confidence with these questions. Colour in 🖉 the bars.

① How do I read correctly from a graph?

② How do I describe the shape of a graph?

③ How do I find a gradient?

① Look at the data in the table.

This data shows how the energy transferred by a resistor changes with time.

To plot this data on a graph you first need to label the axes.

Time in s	Energy transferred in J
4	90
8	200
12	310
16	350

The horizontal and vertical axes on this graph have been partly labelled.

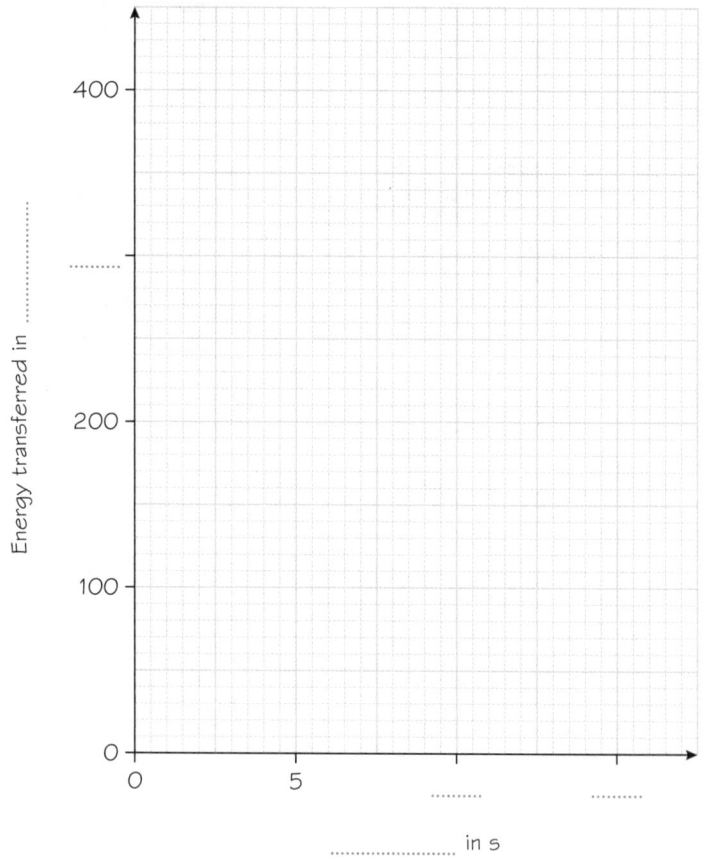

a Complete the labels for the axes using information from the table.

b Complete this sentence:

Each label has the name of the quantity and its ....

Next you need to add scales to the axes. Each square must be worth the same amount.

② **a** Complete the gap on the energy axis.

b How much is each big square worth? ....

③ Complete the gaps along the x-axis to add the scale.

④ Now use the values from the table to plot the points on the graph.

⑤ Draw a line of best fit.

A line of best fit will pass through, or very near, all of the points, except any anomalous points (points which do not fit the pattern).

1 How do I read correctly from a graph?

In your exam you will need to use scientific information presented in the form of a graph. This page will help you to read values from graphs of different types and with different scales.

The **scale** on the axis shows how much each square is 'worth'.

Here is part of a graph.

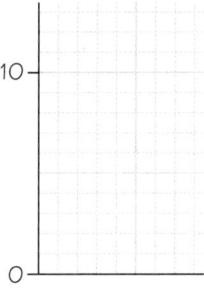

1 How many small squares are there between the 0 and 10 marks? 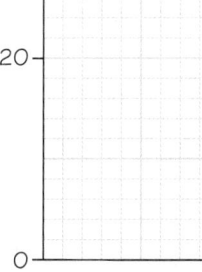 ....

2 How much is each small square worth on the scale? ....

Ten squares are worth 10 units.

Here is part of another graph.

3 How many small squares are there between the 0 and 20 marks? ....

4 How much is each small square worth on the scale? ....

To read from a point on the graph, draw a line across and down until you meet each axis.

5 Complete the sentence for the graph below.

The point is where the length value = ... cm and the

force value = ... N

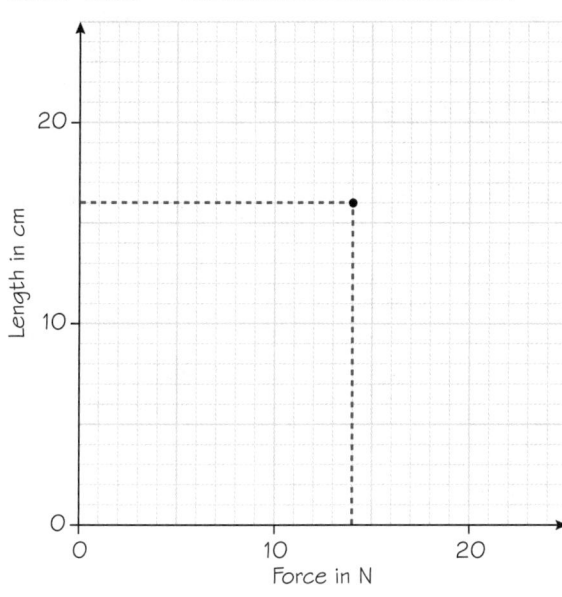

2 How do I describe the shape of the graph?

Marks can be lost when describing graphs if not enough detail is given. This page will help you to work out what to say when asked to describe the shape of a graph.

Look at the example graphs below. They are called **sketch graphs** because they do not have points plotted: they only show the shape.

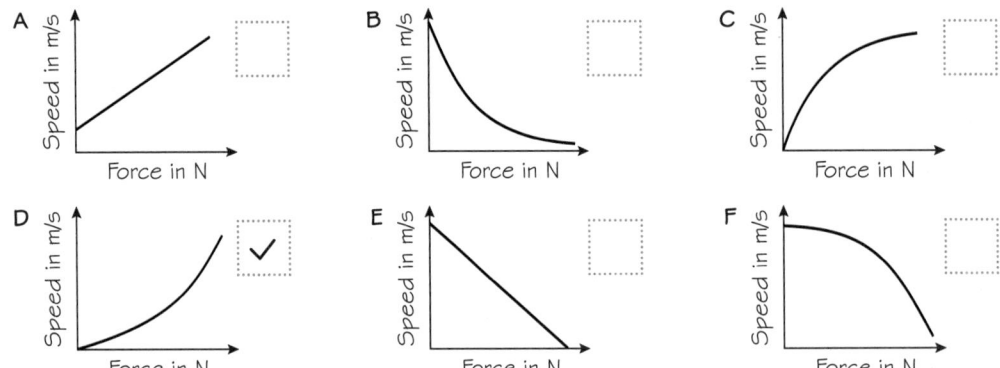

1. **Graph D** matches the statement: As force increases, speed also increases.

   Tick ⊘ any other graphs that show this relationship.

To describe the shape fully, you have to say how these three graphs are different. Do this by describing how the steepness changes.

2. Look again at the three graphs you ticked in ①. Draw ⊘ lines to link the description to the shape of each curve. Write ⊘ the letter of the graph in the boxes on the right.

Description	Shape	Graph
Decreasing rate	A straight line	
Constant rate	A curve that gets steeper	
Increasing rate	A curve that gets flatter	

3. Which graph shows the relationship stated below? ⊘ ....

   > As the force increases, the speed increases at a constant rate.

4. Now describe ⊘ these graphs in the same way. The first description has been started for you.

   a Graph B: *As the force increases, the speed* ....

   b Graph C: ....

   c Graph E: ....

   d Graph F: ....

③ How do I find a gradient?

For many graphs in physics, the steepness or gradient of the line is an important quantity. This page will help you to find the gradient of a section of a graph.

On a **distance–time graph**, the gradient is important: it tells you the speed of the journey. The steeper the line, the higher the gradient so the faster the speed:

gradient = speed

To calculate the speed / gradient at C on the graph we need to use the speed formula we already know.

$$speed = \frac{distance}{time}$$
$$= \frac{160\,m}{80\,s} = 2\,m/s$$

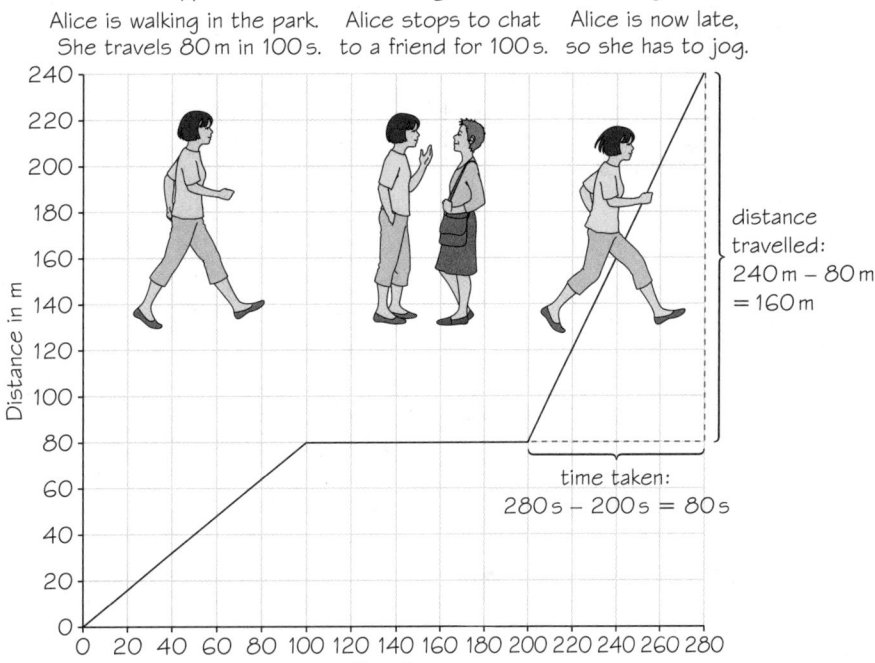

A — Alice is walking in the park. She travels 80 m in 100 s.

B — Alice stops to chat to a friend for 100 s.

C — Alice is now late, so she has to jog.

distance travelled: 240 m – 80 m = 160 m

time taken: 280 s – 200 s = 80 s

① Now look at the first 100 seconds of the journey in the graph above. This is a straight line. Complete ✎ the calculation to find the gradient of the straight line.

$$Gradient = \frac{\text{vertical distance between the start and end points}}{\text{horizontal distance between the same two points}}$$

$$= \frac{80\,m - \boxed{}\,m}{\boxed{}\,s - 0\,s} = \frac{\boxed{}\,m}{\boxed{}\,s} = \dots\dots m/s$$

Here is a **speed–time graph**. This time the gradient is the acceleration; a steeper line means a larger acceleration.

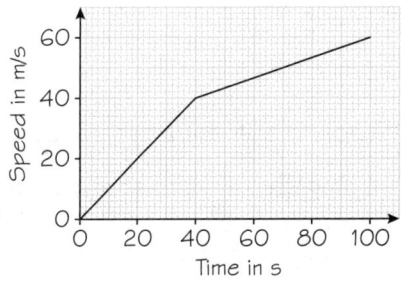

Two students have tried to calculate the acceleration between 40 s and 100 s.

A
$$Gradient = \frac{\text{vertical distance between the start and end points}}{\text{horizontal distance between the same two points}} = \frac{60}{100} = 0.6\,m/s^2$$

B
$$Gradient = \frac{\text{vertical distance between the start and end points}}{\text{horizontal distance between the same two points}} = \frac{47 - 43\,m/s}{80 - 60\,s} = \frac{4\,m/s}{20\,s} = 0.2\,m/s^2$$

② The students have made different mistakes. Explain ✎ what each student has done wrong.

....

....

③ On paper, calculate the correct gradient. ✎

Physics

Sample response

Remember to look for clues in the question. Make good use of your knowledge from other areas of physics.

Exam-style question

1   A student is investigating the relationship between the force on a spring and its extension.

The student's data is shown plotted as a graph in **Figure 1**.

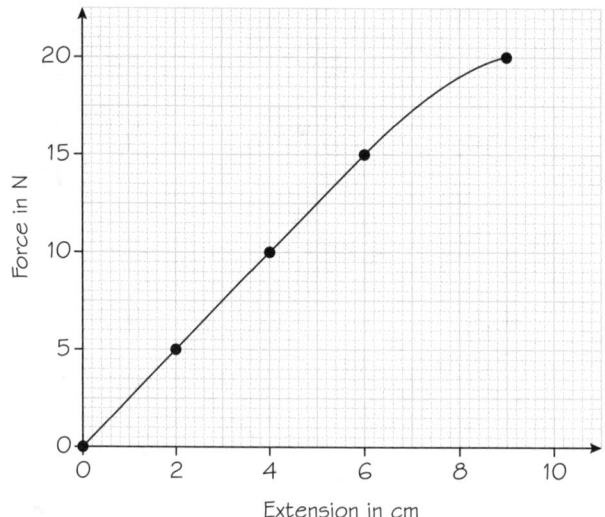

Figure 1

1.1   Describe the relationship shown by the graph.

.... (2 marks)

1.2   Use **Figure 1** to find the extension of the spring when the force is 12 N.

.... (2 marks)

1.3   Use **Figure 1** to find the spring constant in N/m.

.... (2 marks)

① Look at this student's answer to part **1.1**.

Complete ✏ the student answer to describe the part of the graph between 15 N and 20 N.

> *The graph is a straight line until the force is 15 N.*
>
> ....
>
> ....

② Another student has read some data from the graph to answer part **1.2**.

> *The extension is 4.4 cm when the force is 12 N.*

Find 12 N on the vertical axis and draw a line across the graph.

a   How much is each small square on the vertical scale worth? ✏ ....

b   Explain ✏ how the student has read the scale incorrectly.

....

....

Your turn!

It is now time to use what you have learned to answer the exam-style question.

Read the exam-style question and answer it using the guided steps below.

Exam-style question

1 A student is investigating the relationship between the force on a spring and its extension.

The student's data is shown plotted as a graph in **Figure 1**.

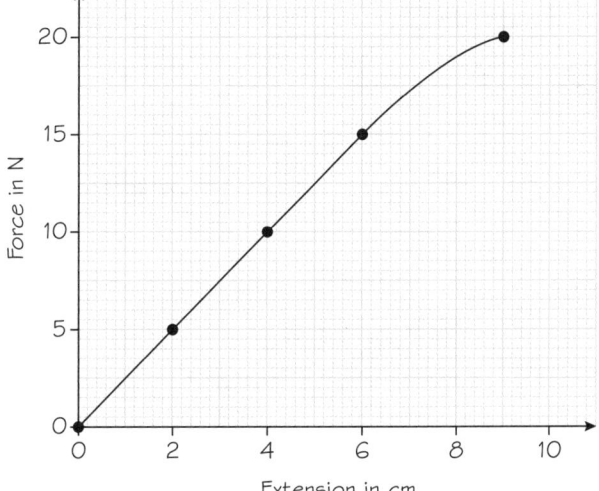

Figure 1

1.1 Describe the relationship shown by the graph.

.... (2 marks)

1.2 Use **Figure 1** to find the extension of the spring when the force is 12 N.

.... (2 marks)

1.3 Use **Figure 1** to find the spring constant in N/m.

.... (2 marks)

1.1 Circle (A) the words that could be used to describe the shape of the graph.

straight line until 15 N	extension increases	force decreases
decreasing gradient	increasing gradient	

1.2 i Draw (✏) a line across the graph where the force is 12 N.

ii Write (✏) the extension of the spring. ....

1.3 i What are the units of extension on the graph? (✏) ....

ii What are the units of the spring constant in the question? (✏) ....

iii What is 6 cm converted into metres? (✏) ....

iv Complete (✏) the gaps in this gradient calculation:

The spring constant is equal to the gradient of this graph when the correct units are used.

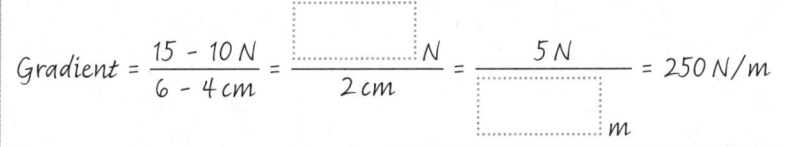

$$Gradient = \frac{15 - 10\ N}{6 - 4\ cm} = \frac{\boxed{}\ N}{2\ cm} = \frac{5\ N}{\boxed{}\ m} = 250\ N/m$$

Need more practice?

In the exam, questions about graph skills could occur as:

- simple standalone questions

- part of a question on any physics topic, especially forces

- part of a question about a practical test.

Have a go at this exam-style question.

Exam-style question

1   The results of an investigation into refraction are shown in the graph below.

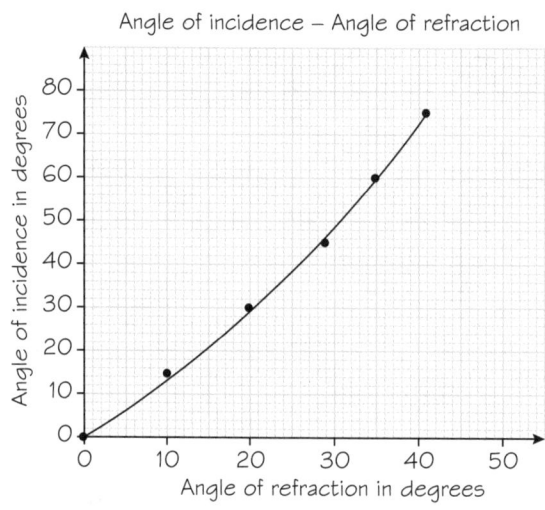

Angle of incidence – Angle of refraction

1.1   Describe the relationship between angle of refraction and angle of incidence.

....

....   (2 marks)

1.2   Use the graph to find the angle of refraction when the angle of incidence is 40 degrees.

....

....   (2 marks)

Boost your grade

Make sure you know how to read graphs to get the information you require as well as being able to draw graphs correctly yourself.

How confident do you feel about each of these **skills?** Colour in the bars.

1 How do I read correctly from a graph?

2 How do I describe the shape of a graph?

3 How do I find a gradient?

⑦ Answering extended response questions

In this unit you will practise planning and writing answers to extended response questions.

In the exam, you will be asked to tackle questions such as the one below.

Exam-style question

1 A student wants to find out the density of a stone. The stone has an irregular shape.

1.1 Describe an investigation the student could complete to find the density of the stone.

Your answer should consider any causes of inaccuracy in the measurements.

.... (6 marks)

You will already have done some work on extended response questions. Before starting the **skills boosts**, rate your confidence with these questions. Colour in 🖉 the bars.

① How do I know what the question is asking me to do?

② How do I organise my answer?

③ How do I choose the right detail/answer the question concisely?

You are likely to meet two kinds of extended response question in physics:

- describing an experiment you have done in class, usually for 6 marks
- explaining a physical process, usually for 4 or 6 marks.

Exam-style question

1   At a concert, lights and loudspeakers are used to produce light and sound waves.

   1.1   Compare the properties of light waves and sound waves.   **(4 marks)**

① The command word in a question is the word that tells you what to do in your answer. Underline Ⓐ the 'command word' in the question above.

The command word is usually the first word of the question itself.

② Circle Ⓐ the two kinds of waves that you are asked to compare.

③ This student has used words in their answer to show they are making a comparison.

> Light waves are transverse whereas sound waves are longitudinal.
> Light travels faster than sound.

Underline Ⓐ the words which are used to show a comparison.

Exam-style question

   2.1   Describe how to find the density of an object.   **(6 marks)**

④ Read these two student answers to the question above.

Student A

> Find the mass of the object by placing it on a balance. Write down the reading.
> Find the volume of the object.
> Divide them to get the density.

Student B

> Find the mass of the object by placing it on a balance.
> Write down the reading. Repeat to check the measurement.
> If the object has a regular shape, find its volume by measuring the sides and calculating the volume using a formula such as:
> volume = length × width × height.
> If the object has an irregular shape, find its volume by a displacement method.
> Use the formula density = $\dfrac{mass}{volume}$ to find the density.

ⓐ Which answer is better? ✎ ....

ⓑ Circle Ⓐ each of the details that makes this a better answer.

⑤ Suggest ✎ one detail Student A could have added to their answer.

....

⑥ Which line from Student B's answer do you think could be improved with more details? ✎

....

1 **How do I know what the question is asking me to do?**

This page will help you decide whether a question is about practical procedures or scientific ideas so you can identify the kind of answer you need to give. You can do this by looking for certain key words or command words in the question. You can usually identify questions that are asking you to describe one of the required practical procedures because they use words in the question that relate to practical methods.

1 Look at the questions below. Tick ✓ the three questions about practical procedures.

Look out for words like plan, design, method, measure, experiment or investigation.

Describe the methods you would use to determine the effectiveness of an insulator.	Plan an experiment to investigate how the force on a spring affects its extension.	Explain the factors that affect whether a driver can stop in time when he sees an obstacle.	Compare the properties of light waves and sound waves.	Explain how you would measure the resistance of a thermistor at different temperatures.

2 Each question will contain a command word. This tells you what you need to include in your answer.

The command word is usually the first word of the question itself which tells you what you are being asked to do.

a Match ✎ each command word to the correct answer content.

Explain		Give facts about two things linked with comparative words like 'more'.
Compare		Give points for and against with a justified conclusion.
Evaluate		Link facts with their consequences.

b Match ✎ the three command words below to the parts of example answers.

		Overall I think nuclear power is the better choice because …
Compare		
Explain		Nuclear power provides a steady power output whereas solar power is unreliable.
Evaluate		Solar power is unreliable because the output depends on weather conditions.

3 Sometimes the question will give you some additional details that have to be included.

Exam style question

1   At a concert, lights and loudspeakers are used to produce light and sound waves.

  1.1   Compare the properties of light waves with sound waves.

    Include ideas about transverse and longitudinal waves in your answer.

    .... **(4 marks)**

Highlight ✎ the part of the question above which tells you that additional information is needed in your answer.

Physics

2 How do I organise my answer?

Once you have decided which information to include in your answer, you need to put your ideas into a logical order. This page will help you practise putting ideas into order.

One of the investigations you will have completed is to measure the speed of water waves in a trough.

(1) A student has written the following statements to describe the experiment.

Put the statements in order by numbering them. 🖉

☐ Repeat the previous steps two more times	☐ At the same time, start the stopwatch
☐ Calculate the mean time taken	☐ Measure the length of a trough
☐ Raise one end a little and drop it	☐ Stop the stopwatch when the wave has travelled to the end of the trough and back again
☐ Use the equation: $$speed = \frac{length\ of\ trough \times 2}{mean\ time}$$ to find the speed of the waves	☐ Half-fill the trough with water

(2) A student has written the following points to answer a question about how a thermistor can be used to measure the temperature in a central heating boiler.

☐ the resistance of the thermistor will be high

☐ Therefore, when the temperature in the boiler is too low

☐ A thermistor has a lower resistance when the temperature is high

☐ This can be detected and used to turn the boiler on

☐ As a result, the current in the circuit will increase

a Highlight 🖉 the conjunctions (connecting words) used in these points.

Conjuctions are terms like 'because', 'so', 'therefore', 'since' and 'due to'.

b Now put the statements in order by numbering them. 🖉

3 How do I choose the right detail/answer the question concisely?

Two of the most common mistakes in extended writing are writing too much irrelevant detail and missing out important details. This page will help you find the right balance.

Here is the start of an exam question.

Exam-style question

1   This question is about the magnetic fields around a straight wire and a solenoid.

(1) A student has completed these diagrams to collect their ideas about wires and solenoids.

circular
depends on distance
depends on current
wire
weaker
straight

depends on current
electromagnet
iron core
solenoid
uniform
stronger
coiled
shaped like a bar magnet

a   Circle (A) pairs of related facts on the two diagrams – for example, 'stronger' and 'weaker'.   Use different colours to show each pair.

b   Cross out (X) any information that isn't needed to answer the question.

(2) Here is the rest of the question.

Exam-style question

1.1   Compare the strength and shape of the magnetic fields around the wire and solenoid.
In your answer, include the factors that affect the strength of the magnetic fields.

a   Circle (A) the command word in the question.

b   Highlight (✐) the additional information you should include.

(3) Look at these sentences from an answer about a practical procedure using a solenoid.
Cross out (X) any irrelevant details in the sentences.

a | Coil the long wire around an HB pencil so it is in the shape of a solenoid.

b | Make sure you have six wires, three red and three black.

c | Go to the bench with the meters and find an ammeter that reads zero when not connected.

Physics

Sample response

To answer an extended response question you need to:

- decide exactly what the question is asking for
- identify the scientific ideas or practical procedures that are relevant
- put the ideas into a logical order
- decide how much detail to write without writing too much.

Now look again at this exam-style task.

Exam-style question

1   A student wants to find out the density of a stone. The stone has an irregular shape.

1.1   Describe an investigation the student could complete to find the density of the stone.

Your answer should consider any causes of inaccuracy in the measurements.

.... **(6 marks)**

Look at this student's answer.

> *First go and collect all your equiptment and get it all together on the table so that it is easy to set up. Make sure your bags and folders are out of the way and if you are a girl you have tied back your hair. Then go to the tap to fill up the measuring cylinder with water. You should use 100 ml but it doesn't matter if it is more or less but you must write down how much is in it so you know how much you have got. Then put the stone in the water and see how much it goes up. Write down the volume of the stone then divide it by the mass and that's the density.*

(1) **a**   Do you think this student has chosen the most **relevant** things to write about? ✐ ....

 b   Explain ✐ your answer.

....

....

 c   Cross out ✐ all the sentences that are not relevant in the answer above.

(2) Explain ✐ whether you think the student has been clear about how to find the volume of the stone.

Have they given the instructions in a **logical order**?

....

....

....

(3) Highlight ✐ any words spelled incorrectly.

Your turn!

It is now time to use what you have learned to answer the exam-style question. Remember to thoroughly read the question and look for clues. Make good use of your knowledge from other areas of physics.

Read the exam-style question and answer it using the guided steps below.

Exam-style question

1   A student wants to find out the density of a stone. The stone has an irregular shape.

   1.1   Describe an investigation the student could complete to find the density of the stone.

   Your answer should consider any causes of inaccuracy in the measurements.

   .... (6 marks)

1 Underline Ⓐ the command word in the question.

2 Highlight ✐ the text that describes the main part of what your answer should cover.

3 Circle Ⓐ the additional detail that you must cover.

4 **a** Which two measurements are needed to calculate the density? ✐

   ....

   b Write down ✐ the formula needed to calculate density from those measurements.

   ....

5 Write down ✐ the steps needed to measure the mass.

   ....

   ....

6 Write down ✐ the steps needed to measure the volume.

   ....

   ....

   ....

   ....

7 **a** Which of the two measurements will be the least accurate? ✐ ....

   b Explain why. ✐

   ....

   ....

8 Now write ✐ your complete response to the exam-style question on a separate sheet of paper.

Physics

Need more practice?

In this section you are going to review your skills and have an opportunity to practise applying them to a new situation.

Have a go at this exam-style question. 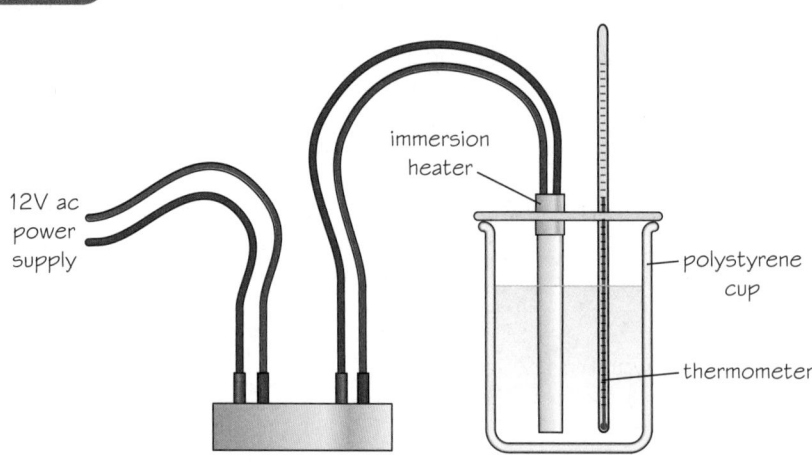 If you run out of space, continue on paper.

Exam-style question

1   A student uses the apparatus shown to investigate how the temperature of water changes after the immersion heater is switched on.

  1.1   Describe how a student could use the apparatus to find the specific heat capacity of water.

  Your answer should consider any cause of inaccuracy in the measurements.

  ....

  ....

  ....

  ....

  ....

  .... (6 marks)

Boost your grade

A more demanding question might ask you to compare two techniques or explain how one physical process differs from another.

How confident do you feel about each of these **skills?** Colour in the bars.

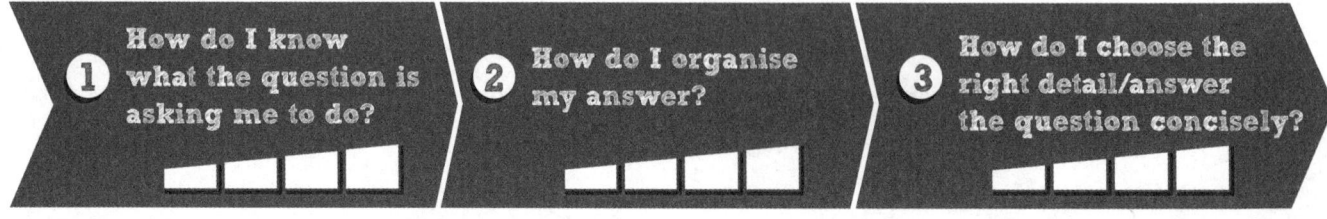

1 How do I know what the question is asking me to do?

2 How do I organise my answer?

3 How do I choose the right detail/answer the question concisely?

Answers

Biology

Unit 1

Page 2

1. Arrow from right to left
2. Water (small blue molecules)
3. Concentration falls
4. A

Page 3

1. B

Exam-style question

1.1 Left-hand side

1.2 Right-hand side

1.3 Carbon dioxide to the left and oxygen to the right

1.4 There is a higher concentration of oxygen on the left. Oxygen molecules will move (diffuse) towards the area with a lower concentration, down the concentration gradient.

Page 4

1. a

	Distilled	10% sucrose	20% sucrose
Change in mass (g)	11 − 10 = 1 g	10 − 10 = 0 g	9 − 10 = −1 g
% change in mass	$\frac{1}{10} \times 100$ $= +10\%$	$\frac{0}{10} \times 100$ $= 0\%$	$\frac{-1}{10} \times 100$ $= -10\%$

b. The piece placed in distilled water

c. This is the same concentration as the inside of the potato cells. There will be no net movement of water molecules in either direction so there is no change in mass.

d. An actual change in mass of 1 g would be a huge change if the initial mass was 2 g, but a very small change if the initial mass was 100 g. The percentage change gives an idea of the size of the change even if you do not know the initial mass. It also allows you to compare the changes in substances with different initial masses.

Page 5

1. transporter protein; energy; against / up

2.

Transport process	Molecules move down the concentration gradient?	Uses energy from respiration?	Molecules move against the concentration gradient?
Diffusion	✓	✗	✗
Osmosis	✓	✗	✗
Active transport	✗	✓	✓

Exam-style question

1.1 The concentration drops to 0.

1.2 Molecules are moved from low concentration to higher concentration, against the concentration gradient. Energy from respiration is used.

Page 6

1. a. The response is a description, not an explanation.

b. Osmosis

c. There is a higher salt concentration in the sea water so water moves out of the plant roots by osmosis.

2. There is a plant absorbs mineral ions by active transport.

They are used to build proteins / cell membranes / for photosynthesis / for respiration.

Page 7

Exam-style question

Substance entering or leaving cell	How the substance crosses the cell membrane
Minerals entering a plant root from the soil	Enter by active transport: energy from respiration is used to move minerals from a low concentration to a higher concentration in the plant
Oxygen entering a liver cell	Enters by diffusion, from a high concentration to a lower concentration
Water entering a plant root from the soil	Enters by osmosis, from a dilute solution to a more concentrated solution
Glucose taken up from the small intestine	Taken up by active transport, using energy from respiration, so that all the glucose is taken up

(1 mark for each correct row)

Page 8

1.1 −10.8 / −7.1 / −5.6 / −2.6 / +6.3
all values correct **(2)**, one or more incorrect **(1)**

1.2 The potato had a higher sugar solution concentration than the surrounding solution **(1)**.

Water moved from the more dilute solution to the more concentrated solution by osmosis **(1)**, so water moved into into the potato, increasing its mass **(1)**.

Unit 2

Page 10

(1) **(a)** C

(b)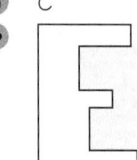

(2) active site; fit; denatured

Page 11

(1) **(a)** substrate; product; faster

(b) 20; 40; 40; 60

(2) **(a)** a protein

(b) amino acids

(c) Proteins are broken down / digested into amino acids by the protein-digesting enzyme (protease).

(d) Rises steeply at first, then more slowly and then levels off

(e) As the time increases, the concentration of amino acids rises quickly at first to approximately 22.5 mg/dm³ at 10 minutes. Then, for the next 10 minutes, the reaction slows down and the concentration increases more slowly, up to 28.0 mg/dm³ at 20 minutes. The graph then begins to level off, reaching a concentration of 30.0 mg/dm³ at 30 minutes.

Page 12

(1) **(a)** At substrate concentration of 2.0 mol/dm³

(b) When the substrate concentration is zero there are no substrate molecules. Therefore, there will be no collisions and the rate of reaction is zero.

As the substrate concentration increases there are more substrate molecules. Therefore the substrate molecules collide with the enzyme molecules more frequently and the reaction rate is higher.

The highest number of collisions is when the substrate concentration is at 2.0 mol/dm³ so this gives the highest rate of reaction.

(2) **(a)** C

(b) A

(3) **(a)** All the starch has been digested, so there are no collisions.

(b) The concentration of starch is decreasing, so there are fewer collisions.

Page 13

(1) **(a)** substrate is hydrogen peroxide; product is oxygen

(b) very little; very slowly; not very often

(c) lots of / more; more often

(d) 40 °C

(e) It falls to zero

(f) The enzyme has denatured: this means the active site has changed shape so the substrate will no longer fit into it.

Page 14

(1) There will be more collisions with enzyme molecules so the rate of reaction will be higher.

(2) 50 °C

(3) At low temperatures, the enzyme and substrate molecules do not have much kinetic energy so they do not collide very often. At 50 °C the molecules have more kinetic energy, so they move faster. There are more collisions between enzyme and substrate molecules, so the rate of reaction is faster. At higher temperatures, the enzyme starts to denature. The shape of the active site changes, so the substrate can no longer fit into it. The rate of reaction slows down and eventually stops altogether.

Page 15

1.1 At pH 2 the rate of digestion is zero **(1)**. As the pH rises, the rate of reaction rises to a maximum of 0.55 s⁻¹ at pH 7 **(1)**. Above pH 7 the rate drops again **(1)**.

1.2 pH 7 **(1)**

1.3 This is the optimum pH, where the active site fits the substrate **(1)**. At higher and lower pH values, the active site changes shape, so the substrate cannot fit into it **(1)**.

1.4 The enzyme was denatured by the low pH **(1)**, so the substrate did not fit the active site **(1)**.

Page 16

1.1 A catalyst is a substance that speeds up the rate of a reaction without itself being used up **(1)**.

1.2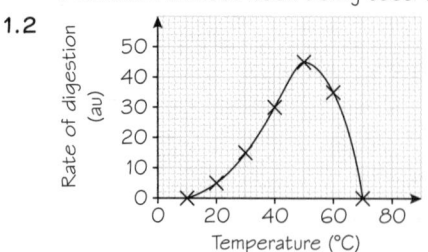

smooth line passing through all points **(1)**

1.3 20–22 **(1)** arbitrary units **(1)**

1.4 The enzyme has denatured, so the active site has changed shape **(1)**. The substrate (starch) no longer fits the active site properly **(1)**.

Unit 3

Page 18

1.

Stage	Correct order
The cell increases in size and increases the number of sub-cellular structures such as ribosomes and mitochondria. DNA replicates to form two copies of each chromosome.	1
The cytoplasm and cell membrane divide to form two identical, daughter cells.	3
A set of chromosomes moves to each end of the cell and the nucleus divides.	2

2.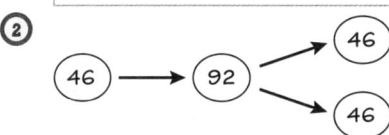

3. a. The genetic information / DNA / chromosomes is copied / doubled / duplicated.

   b. The cell membrane and cytoplasm divide to form two new identical daughter cells.

4.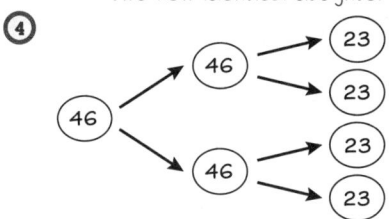

Page 19

1. a.

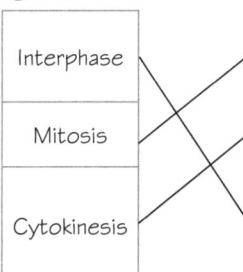

Interphase		One set of chromosomes is pulled to each end of the cell, and the nucleus divides.
Mitosis		The cell membrane and cytoplasm divide to form two new cells.
Cytokinesis		The cell increases in size and produces more ribosomes and mitochondria. The cell also makes a complete copy of the DNA.

   b. i 360 minutes

   ii 1 minute

   iii

	Angle (°)	Time in minutes	Time in hours and minutes
Interphase	320	320 min	5 h 20 min
Mitosis	20	20 min	0 h 20 min
Cytokinesis	20	20 min	0 h 20 min

2. a. 20 chromosomes / double / twice as many

   b. 10 chromosomes (in each daughter cell)

3. a. X

   b. Y

Page 20

1. nucleus 3   gene 1   chromosome 2

2. a. mitosis

   b. It replicates / doubles / is copied / duplicates / is cloned.

   c. 46

---

d. For growth / repair / replacement of damaged cells

3. Completed diagram should be exactly the same as the original.

Page 21

1. B

2. testes / ovaries

3. BF Bf bF bf

4. halves; non-identical; diploid; haploid; sexual; increases

Page 22

1. There are two ticks in the first row, so the student would not get a mark for this row.

   There is no tick in the second row, so the student would not get a mark for that part of the question.

2. Could also add:

   Each of the cells is different.

   Each cell has half the number of chromosomes of the original cell.

3. Growth requires identical copies of cells to be made. It is mitosis that produces identical cells.

Page 23

Exam-style question

1.1 (1 mark for each correct row)

Feature	Mitosis or meiosis?
Production of eggs	Meiosis
A lizard growing a new tail	Mitosis
Production of pollen in a flower	Meiosis
Cells replaced on the skin to heal a cut	Mitosis

1.2 A man: testes (1)

   A woman: ovaries (1)

1.3 Mitosis produces identical cells; meiosis produces non-identical cells (1).

   Mitosis produces two daughter cells; meiosis produces four gamete cells (1).

   Mitosis is for asexual reproduction, growth and repair; meiosis is for sexual reproduction (1).

   Mitosis occurs in all body cells except for the testes and ovaries; meiosis occurs only in the testes and ovaries (1).

   (any two points for 2 marks)

Page 24

Exam-style question

1.1 meiosis (1)

1.2 mitosis (1)

1.3 zygote / fertilised egg (1)

1.4 4 pg (half) (1)

Unit 4

Page 26

1. a. Independent variable: temperature

   Dependent variable: height of bubbles

   b. Experiment is carried out in a water bath at set temperatures

c Same volume of same concentration used in all stages of the experiment; all other volumes the same in all stages

d Always use 2% hydrogen peroxide; keep all volumes the same in all stages

② Temperature (°C)   Height of bubbles (mm)

Page 27

① a substrate concentration

b pH

③ a clarity of apple juice

b length of a potato chip

③ a Two from: temperature, pH, concentration of CO_2, type of plant

b Two from: temperature, concentration of all chemicals

④ a speed of air movement

b rate of water loss

c Three from: temperature, size of plant, type of plant, light intensity, availability of water to the plant's roots

Page 28

① moving the light

② counting the number of bubbles released per minute

③ a Use a stop watch to time a set amount of time (e.g. 1 minute) and count all the bubbles released in that time.

b Add the same mass of sodium hydrogencarbonate.

④ pondweed; 5 / a few; minute; 3; 4; 20; 30

Page 29

① Temperature (°C)   Time taken to digest protein (seconds)

② a light intensity

b Two from: independent variable is in right-hand column; dependent variable is in left-hand column; no units are given

③ a temperature

b rate of water loss

c

	Rate of water loss from shoot (g)			
Temperature (°C)	Trial 1	Trial 2	Trial 3	mean
10				
20				
30				

Page 30

① They could have stated that it was concentration of food dye.

② A unit for the spread of dye

③ Two

Page 31

Exam-style question

1.1 The distance the ruler drops before it is caught (1)

1.2 Two from: same ruler, same starting point (0 cm mark held between thumb and forefinger), same level of distraction in room around subject, same people being tested (1 mark for each correct point)

1.3 Different people were tested at different times of day (1); the same person should be tested at different times to see how reaction time is affected by time of day (1).

Page 32

Exam-style question

1.1 shade / light intensity; soil moisture; soil pH; mineral concentration in soil
all answers correct (3); one incorrect (2); two incorrect (1); more than two incorrect (0)

1.2 in shade = 2.7 (1); in sun = 5.7 (1)

1.3 Place the quadrat three times in the sun and three times in the shade (1); state size of quadrat (1); state how position of quadrat was selected (this should be randomly selected) (1).

Unit 5

Page 34

① a How the percentage cover of reeds changes as distance from the pond increases

b B

② a 13%

b 7 m

③ The percentage cover of reeds is highest close to the pond and decreases as distance from the pond increases. The soil is wetter near the pond. The data show that reed plants prefer wetter conditions.

Page 35

① a 16 cm³/s

b 14 cm³/s

c 4 mg/dm³

② a values up to 8 mg/dm³

b curved, becoming less steep

c concentration; increases; 12; straight; constant; curved; 16; 18

Page 36

① a distance of lamp from plant, in cm

b rate of photosynthesis, in bubbles per minute

c decreases

d decreases

e curve; falls; 10 cm; less; 35

f quickly; 6; slowly; 35

Page 37

① the effect of changing light intensity on the rate of photosynthesis

② A

③ The light provides the energy for photosynthesis.

④ photosynthesis; energy; intensity; decreases; decreases; photosynthesis; 35

Page 38

① the units (%)

② The chances of dying decrease from 20% to 14%; this is not halving.

③ 2 marks. The student has described the shape of the curve [1 mark], and stated some correct data values from the graph [1 mark]. However, the student has mistaken coronary heart disease for cancer, and has not correctly explained what the graph shows.

Page 39

Exam-style question

1.1 For the first 25 hours the graph is a horizontal line, which shows that the bean root is not growing **(1)**. From 25 hours to 100 hours the graph is a curve, rising steeply **(1)**. This shows that the root grows quickly, from 0 mm at 25 hours to 33 mm at 100 hours **(1)**.

1.2 50 mm **(1)**

1.3 The seed must absorb water from the surroundings **(1)**, this will activate the enzymes that start to digest its stored food. This allows growth to start **(1)**.

Page 40

Exam-style question

1.1 As the blood alcohol level increases, the risk of an accident increases. The graph is a curve that rises slowly at first **(1)**. This means that for blood alcohol levels up to about 60 mg per 100 cm^3, the risk of an accident increases slowly **(1)**. For blood alcohol levels from about 60 mg per 100 cm^3 to 200 mg per 100 cm^3, the curve rises more steeply. This means that as the alcohol level rises above 60 mg per 100 cm^3 blood, the risk increases more quickly **(1)**.

1.2 4 **(1)**

1.3 120 mg per 100cm^3 blood **(1)**

Unit 6

Page 42

1 **a** 30 mm

    b $\dfrac{30}{6}$

    c ×5

2 **a** 20 cm^3 produced in 5 s

    b 4 cm^3/s in 1 s

    c 4 cm^3/s

3 **a** $\dfrac{4}{15}$

    b $\dfrac{4}{15}$ = 0.266666

    c 0.266666 × 100 = 26.6666%
       = 26.7% (to 1 d.p.)

Page 43

1 **a** length of object = 0.04 mm, length of image = 80 mm

    b $\dfrac{80}{0.04}$ = 2000

    b magnification = ×2000

2 **a** actual size = $\dfrac{\text{image size}}{\text{magnification}}$

    b actual size = $\dfrac{5 \text{ mm}}{200}$

    c actual size = 0.025 mm

Page 44

1 **a** 24 cm^3 produced in 3 minutes

    b divide by 3: 8 cm^3 in 1 min

    c rate of reaction = 8 cm^3/minute

2 **a** 8 mm/h

    b 30 − 24 = 6 mm in 1 hour
       rate = 6 mm/h

3 **a** 12 mm growth in 50 hours

    b $\dfrac{12 \text{ mm}}{50 \text{ hours}}$

    c rate of growth = 0.24 mm/h

Page 45

1 $\dfrac{24}{80}$ = 0.3 = 30.0%

2 $\dfrac{29\,435}{45\,692}$ = 0.6442 = 64.42% = 64.4% (to 1 d.p.)

3 **a** 385 000 − 345 000 = 40 000

    b original value = 345 000

    c percentage change = $\dfrac{40\,000}{345\,000}$ × 100
       = 11.594% = 11.6% (to 1 d.p.)

4 actual change = 352 − 324 = 28

original value = 324 tonnes

percentage increase = $\dfrac{28}{324}$ × 100 = 8.642%
= 8.6% (to 1 d.p.)

Page 46

1 **a** Measure the length of the scale provided for 1 μm and use this to estimate the actual length of the bacterium.

    b image = 12 mm = 12 000 μm

    actual length = 1 μm

    magnification = $\dfrac{12\,000}{1}$ = ×12 000

2 **a** length of time for growth = 2017 − 1832
    = 185 years

    b You cannot have 0.108 108 of a person / they have not rounded to a sensible degree of accuracy.

Page 47

Exam-style question

1.1 10 cm **(1)**

1.2 length at 120 hours = 10 cm, length at 150 hours = 17 cm, growth = 7 cm

rate of growth = 7 cm in 30 hours = $\dfrac{7}{30}$ **(1)**

= 0.233 333 = 0.23 cm/h (to 2 d.p.) **(1)**

1.3 length at 120 hours = 10 cm, length at 180 hours = 24 cm, actual change = 14 cm **(1)**

percentage increase = $\dfrac{14}{10}$ × 100 **(1)**

= 140% **(1)**

Page 48

1.1 length of guard cell on image = 8 mm,

magnification = $\dfrac{8}{0.04}$ **(1)**

= ×200 **(1)**

1.2 Rate of photosynthesis = $\dfrac{2.4}{8}$ **(1)**

= 0.3 cm^3/h **(1)**

1.3 actual change = 7.2 − 2.4 = 4.8

original value = 2.4 **(1)**

percentage increase = $\dfrac{4.8}{2.4}$ × 100 **(1)**

= 200% **(1)**

Unit 7

Page 50

① ⓐ Explain

ⓑ B

ⓒ so they move more quickly and collide more often

② ⓐ Describe the similarities and / or differences between things.

ⓑ Any three from: <u>better</u> adapted; <u>more likely</u> to survive; <u>more likely</u> to have the good adaptations; <u>much quicker</u> process as the non-adapted individuals do not breed.

Page 51

① ⓐ Describe

ⓑ Explain

ⓒ Draw

ⓓ Calculate

② ⓐ Explain

ⓑ coronary heart disease

ⓒ how being obese and smoking cigarettes increases

ⓓ smoke damages the artery wall; high blood pressure can damage the artery wall; fats build up in the artery

③ ⓐ Describe

ⓑ enzyme

ⓒ test a range of tissues for catalase activity

ⓓ decomposition of hydrogen peroxide to water and oxygen gas; gas can be collected by displacing water in an upturned test tube

Page 52

① Explain

② pH and temperature

③ ⓐ The active site is denatured by high temperature

At low temperatures enzyme activity is low

As temperature rises activity increases

This is because the enzyme molecules have more kinetic energy and collide more often

ⓑ At very high or low pH most enzymes are inactive

Extremes of pH will alter the shape of the active site

Enzymes work best at a particular pH called the optimum pH

④ ⓐ The active site is denatured by high temperature (4)

At low temperatures enzyme activity is low (1)

As temperature rises activity increases (2)

This is because the enzyme molecules have more kinetic energy and collide more often (3)

ⓑ At very high or low pH most enzymes are inactive (2)

Extremes of pH will alter the shape of the active site (3)

Enzymes work best at a particular pH called the optimum pH (1)

Page 53

① Make something clear, or state the reasons for something happening

② ⓐ C, D, E, F, G

ⓑ C = 2, D = 1, E = 5, F = 4, G = 3

Page 54

① Yes

② The coolness stops the microbes growing but also keeps the meat fresh without affecting the taste.

③ Yes

④ No: the response does not cover the possible use of bleach.

⑤ Yes it is logical. However, it could be improved by writing shorter paragraphs on each method or numbering the statements.

Page 55

① Evaluate

② Coronary heart disease

③

Treatment	Benefit	Risk
Drugs such as statins	Reduce blood cholesterol and deposition	Some side effects possible
Surgical – stents fitted	Not invasive surgery, holds the arteries open	Risk of infection from surgery, need to take anti-clotting drugs
Heart transplant	Improved health	Invasive surgery, risk of infection, finding a donor can be hard

④ Put the benefits and risks of each treatment together.

⑤ CHD can be treated by taking drugs such as statins, which help to reduce blood cholesterol levels. This reduces the risk of fat building up in the arteries and is relatively cheap. However, these medicines may cause side-effects, such as allergies or unexpected reactions in the cells.

If the arteries have been narrowed, this can cause conditions such as angina. In these cases, a stent can be fitted. This is a small mesh tube that holds the artery open. It is fitted via a small incision in the groin. A stent gives immediate relief and there is little recovery needed after the operation. However, there is always a risk of infection or blood clots forming. The patient will need to take anti-clotting drugs.

If the heart actually fails , a heart transplant will be needed. This could restore an active lifestyle but there are risks associated with open chest surgery, such as infection, and rejection is possible. The main difficulty is finding a suitable donor.

Page 56

Exam-style question

1.1 Divide food into four parts. Crushing the food will help it to dissolve.

First sample: Dissolve the sample in water and add Benedict's reagent, then heat. A change in colour from blue to yellow / green / red shows sugars are present **(1)**. Wear goggles while heating **(1)**.

Second sample: Dissolve the sample in water and add iodine. A colour change from yellow-brown to blue-black shows starch is present **(1)**.

Third sample: Dissolve the sample in water and add biuret reagent. A colour change from blue to purple shows protein present **(1)**.

Fourth sample: Dissolve the sample in alcohol, filter, and add water to the clear filtrate. A change from clear to cloudy shows fat is present **(1)**. Keep alcohol away from flame **(1)**.

Chemistry

Unit 1

Page 58

1.
   a. methane and oxygen
   b. carbon monoxide and water
   c. the = should be an arrow $\longrightarrow$

2. CO_2

3. Elements: N_2, O_2, Ar, Ne, Kr, Xe, He
   Compounds: CO_2, CH_4

4. aq

Page 59

1. crystals, dilute, solution, gas
2. octane + oxygen $\longrightarrow$ carbon dioxide + water
3. potassium carbonate + nitric acid $\longrightarrow$ potassium nitrate + carbon dioxide + water
4.
   a. iron
   b. iron oxide

Page 60

1.

gas	aq
liquid	s
solid	l
solution in water	g

2. magnesium (s) + hydrochloric acid (aq) $\longrightarrow$ magnesium chloride (aq) + hydrogen (g)

3. The acid is a <u>colourless</u> solution and is placed in a <u>beaker</u>. The copper carbonate, a green <u>powder</u>, is added and stirred with a <u>glass rod</u>. The copper carbonate <u>disappears</u> and a <u>blue solution</u> appears. <u>Bubbles</u> of carbon dioxide are seen.

Page 61

1.
   a.

Formula	Name of substance	Number of atoms of each element in the formula
Fe	iron	Fe = 1
Cl_2	chlorine	Cl = 2
$FeCl_3$	iron chloride	Fe = 1 Cl = 3

   b.

	Reactants side of equation	Products side of equation
Fe	2 × 1 = 2	2 × 1 = 2
Cl	3 × 2 = 6	2 × 3 = 6

2. write a 2 in front of HCl: $H_2 + Cl_2 \rightarrow 2HCl$
3. 2

Page 62

1.
   a.
   i. The reactants and the products are on the wrong sides of the equation because mercury oxide is decomposing. Mercury oxide is the starting material (reactant).
   ii. The word 'oxygen' must be used, because this is a word equation.
   iii. An arrow must be used, not an =
   b. mercury oxide $\longrightarrow$ mercury + oxygen

2.
   a. H = hydrogen, 2 atoms   S = sulfur, 1 atom
      O = oxygen, 4 atoms
   b.

	Reactants side of equation	Products side of equation
Li	2 × 1 = 2	2
H	2	2
S	1	1
O	4	4

   c. is; the same number

Page 63

Exam-style question

1.1 sodium + water $\longrightarrow$ sodium hydroxide + hydrogen
   left hand side **(1)**, *right hand side* **(1)**

1.2 i sodium = Na, hydrogen = H

   ii

	Reactants side of equation	Products side of equation
Na	?	2 × 1 = 2
H	2 × 2 = 4	2 × 1 = 2
O	2 × 1 = 2	2 × 1 = 2

   formulae of Na and H_2 **(1)**, *balancing number 2 in front of Na* **(1)**

   iii $2Na + 2H_2O \rightarrow 2NaOH + H_2$

1.3 hydrogen (g), sodium (s), sodium hydroxide (aq), water (l) **(1)** *for each symbol*

Page 64

Exam-style question

1.1 zinc + nitric acid $\longrightarrow$ zinc nitrate + hydrogen
   left hand side **(1)**, *right hand side* **(1)**

1.2 silver nitrate + hydrochloric acid $\longrightarrow$ silver chloride + nitric acid
   left hand side **(1)**, *right hand side* **(1)**

1.3 zinc (s), nitric acid (aq), zinc nitrate (aq), hydrogen (g), silver nitrate (aq), hydrochloric acid (aq), silver chloride (s)
   all solids = s **(1)**, *all solutions = aq* **(1)**, *hydrogen = g* **(1)**

2.1 $Mg + \underline{2}HCl \longrightarrow MgCl_2 + H_2$
   There are now 1 Mg, 2 H and 2 Cl on each side.

2.2 $2C + O_2 \longrightarrow \underline{2}CO$

2.3 $CH_4 + \underline{2}O_2 \longrightarrow CO_2 + \underline{2}H_2O$

Unit 2

Page 66

1.
   a.

hydrochloric acid	HNO_3
nitric acid	H_2SO_4
sulfuric acid	HCl

   b.

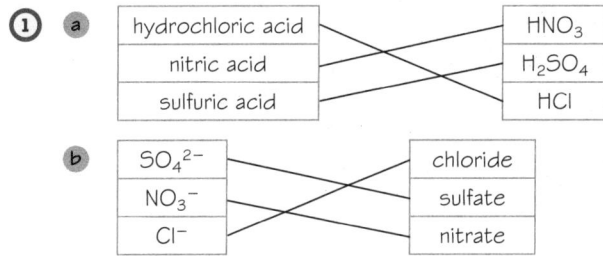

SO_4^{2-}	chloride
NO_3^-	sulfate
Cl^-	nitrate

(2) acid + metal ⟶ salt + <u>hydrogen</u>

acid + base/alkali ⟶ salt + <u>water</u>

acid + metal carbonate ⟶ salt + <u>water</u>
<u>+ carbon dioxide</u>

(3) evaporating basin, tripod, gauze, Bunsen burner

Page 67

(1) **a** potassium carbonate

b potassium chloride

(2) Stage 1: C, F, A

Stage 2: D

Stage 3: B, E, G

(3) filter funnel, filter paper, flask, beaker

The mixture is potassium chloride solution and potassium carbonate.

The residue is potassium carbonate.

The filtrate is potassium chloride solution.

Page 68

(1) **a** Base: zinc oxide;   Salt: zinc sulfate

b the reaction is too slow at room temperature.

c so that all of the acid is neutralised

d

heat with a water bath until saturated	✓
heat with an electric heater until saturated	✓
leave in a warm place	✓

(2) **a** KCl

b Na_2O

c $MgBr_2$

Page 69

(1) **a** Wear safety glasses / goggles

b A measuring cylinder

(2) **a** Stir with a glass rod.

b Some solid will remain on the bottom of the beaker.

(3) **a** Filtration

b Evaporating basin

(4) **a** Powder

b A water bath / electric heater

Page 70

(1) **a** Pour – Measure

test tube – beaker

spatula – glass rod

a residue – an excess

Pour off – Filter

all – about half

writing paper – filter paper

b white crystals

(2) **a** hydrogen

b magnesium + hydrochloric acid ⟶ magnesium chloride + hydrogen

Page 71

Exam-style question

1.1 **Step 1:** Measure some sulfuric acid into a beaker. Warm the acid using a Bunsen burner. **(1)** Add a spatula of copper oxide and stir with a glass rod until the copper oxide has disappeared. **(1)** Keep adding copper oxide and stirring until an excess remains. **(1)**

Step 2: Filter the mixture (pour it through a funnel lined with filter paper) and collect the filtrate in an evaporating basin. **(1)**

1.2 **Step 1:** It is not necessary to filter the solution as it is pure. **(1)**

Step 2: Use an evaporating basin, not a beaker. **(1)**

Step 3: Heat with a water bath or electric heater. **(1)**

Step 4: Stop heating when the solution is saturated, then leave in a warm place so crystals form. **(1)**

Page 72

Exam-style question

1.1 Wear eye protection. **(1)** Use a measuring cylinder to measure 25 cm^3 nitric acid. Place the acid in a 250 cm^3 beaker. Warm the acid using a Bunsen burner, tripod and gauze. **(1)** Use a spatula to add some calcium carbonate to the acid, then stir with a stirring rod until the solid has disappeared. **(1)** Keep adding more calcium carbonate and stirring until an excess remains. Use a filter funnel lined with filter paper to filter the mixture, collecting the solution in an evaporating basin. **(1)** Place the evaporating basin in a water bath and heat until the solution is saturated. **(1)** Then leave in a warm place to allow crystals to form. **(1)**

1.2 $CaCO_3 + 2HNO_3 \longrightarrow Ca(NO_3)_2 + H_2O + CO_2$ **(1)**

Unit 3

Page 74

(1) **a** giant metallic lattice

b giant ionic lattice

c giant covalent lattice

d small molecules

(2) **a** CH_4

b 10

c 4

d iii

Page 75

(1) **a** First shell (inner shell): 2

Second shell: 8

Third shell (outer shell): 7

b 2,8,7

c Chlorine is in group 7 and it has seven electrons in its outer shell.

d

e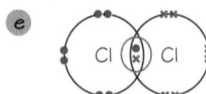

f share; covalent; 8

g i   water

ii   oxygen

iii nitrogen

Page 76

(1) **a** 8

b 2

(2) a 6, 6, two

b

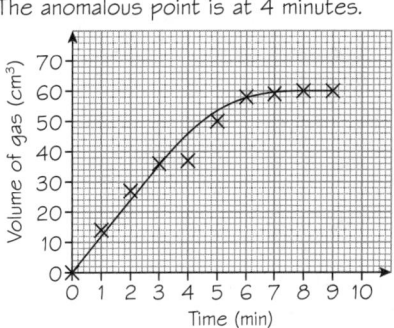

c

Page 77

(1) Diamond: solid

Hydrogen chloride: gas

(2) C

(3) A

(4) B

Page 78

(1) a 5, 3

b

c triple

(2) error 1: the inner shells are shown – the question says show the outer shells only

error 2: there are four shared electrons – there should only be two

error 3: only dots have been used – the question says use dots for one atom and crosses for the other atom

(3) −259 °C

Page 79

Exam-style question

1.1 Each hydrogen atom shares its electron with the nitrogen atom, which also shares one electron with each hydrogen atom. (1) This is a covalent bond. (1)

1.2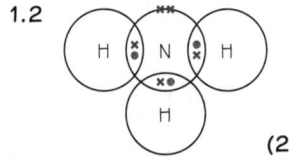

(2)

1.3 Only the intermolecular forces are overcome when ammonia boils. (1) The intermolecular forces are weak so only a little energy is needed to overcome them. (1) This is why the boiling point is low. (1)

1.4 No. (1) There are no free charged particles to carry the current. (1)

Page 80

Exam-style question

1.1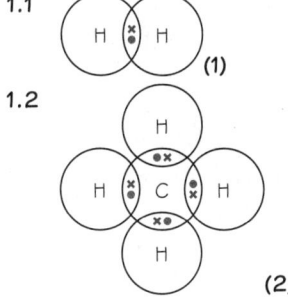

(1)

1.2

(2)

1.3 Small molecules: Z (1)

Giant covalent lattice: X (1)

Unit 4

Page 82

(1) a decreases

b increases

c increases

d increases

e does not change

(2) a The anomalous point is at 4 minutes.

b

c The reaction has finished after 8 minutes, because no more gas is formed after this time. The amount of gas formed stays at 60 cm³ after 8 minutes.

Page 83

(1) a

using a thicker cross when repeating an experiment		the time recorded will be shorter
starting the stop clock after swirling		no effect on the time
using acid of a higher concentration		the time recorded will be longer

No errors link to 'no effect on the time'.

b measuring cylinder

(2) a so the total volume of the mixture is the same

b Experiment 3 (15 cm³ sodium thiosulfate + 15 cm³ water).

Page 84

(1) a C: The reaction is fastest here.

D: The reaction is slowing down here as the hydrogen peroxide is used up.

A: The reaction has stopped here.

E: The total volume of oxygen formed is shown here.

B: The time for the reaction to end is shown here.

b 75 cm³ (Accept answers in the range 73–77 cm³)

c i greatest

ii used up

d

Page 85

1. a gain
   b higher
   c faster
   d more

2. The rate of reaction will increase.

3. more; faster

4. a Increase the temperature
   b The rate of reaction will increase.
   c More particles will have the activation energy (the energy needed to react).

Page 86

1. The answer should refer more clearly to the data in the table.

   The answer should mention P, Q and R and say whether each one is a catalyst.

   The reason given for choosing Q does not explain properly why Q is a catalyst.

   Q is not a catalyst.

   Two of the substances are catalysts.

2. a break lumps of S up (into a powder)
      increase the temperature
   b lower; larger

Page 87

Exam-style question

1.1 Use a measuring cylinder to measure out the hydrochloric acid. Add a known mass of marble chips. **(1)** Measure the volume of gas given off every minute for a certain time. **(1)**

   Repeat the experiment but increase the temperature of the acid. Measure out the same volume of acid into a flask. Place the flask in a water bath to increase the temperature. **(1)** Add the same mass of marble chips and measure the gas given off for the same amount of time. **(1)**

1.2 X shows the results at 40 °C. **(1)** This is the graph with the steeper gradient so the rate of reaction is faster. The reaction is faster at the higher of the two temperatures. **(1)**

1.3 At the higher temperature the particles have more energy, so they move faster. **(1)** They will collide more frequently. **(1)**

   At the higher temperature a larger proportion of the particles will have the activation energy, so a larger proportion of the collisions will be successful (i.e. will result in a reaction). **(1)**

Page 88

Exam-style question

1.1 At a higher pressure there are more particles in the same space. **(1)** This means the particles will collide more frequently, **(1)** so the rate of reaction will increase. **(1)**

1.2 B **(1)**

Unit 5

Page 90

1. a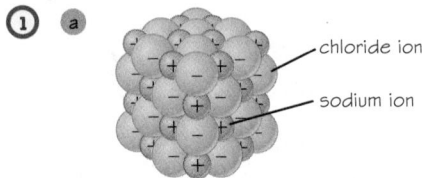

   chloride ion
   sodium ion

b Metal ions are always positive, so the ions with the + must be the metal (sodium). Non-metal ions are always negative, so the ions with the − must be the non-metal (chloride).

2.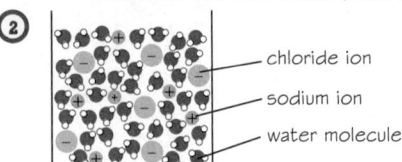

   chloride ion
   sodium ion
   water molecule

3. solid = no; melted = yes; dissolved = yes

Page 91

1. gained; oxidised

2. a i Fe_2O_3: 2 Fe atoms, 3 O atoms
      2Fe: 2 Fe atoms, no O atoms
     ii lost; reduced
   b i 3CO: 3 C atoms, 3 O atoms
      CO_2: 3 C atoms, 6 O atoms
     ii gained; oxidised

3. a, b In each equation, $2e^-$ should have been circled.

Page 92

1. a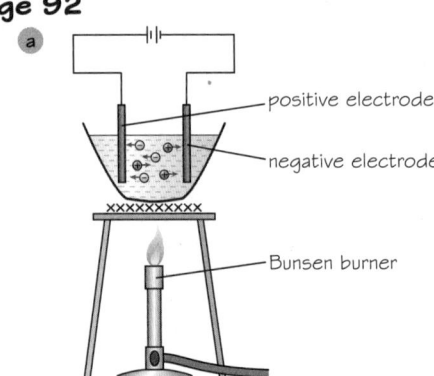

   positive electrode
   negative electrode
   Bunsen burner

   b iii to melt the compound so the ions are free to move

   c Underlined:
   iodide ions; I^-, $2I^- \rightarrow I_2 + 2e^-$; oxidation
   Circled:
   potassium ions; K^+; $K^+ + e^- \rightarrow K$; reduction

   d potassium iodide $\rightarrow$ potassium + iodine

Page 93

1. a K^+; I^-; H^+; OH^-
   b negative; positive

2. hydrogen; iodine

3. a copper
   b oxygen
   c a

Page 94

1. cells / battery / power source

2. The solid is corrosive. Wear gloves when handling the zinc chloride.

3. heat it until it melts

4. silver-coloured

5. chlorine

6. reduced; oxidised

Page 95

Exam-style question

1.1 A giant ionic lattice **(1)** – a lattice of regularly arranged ions. **(1)**

1.2 The ions cannot move. **(1)**

1.3 Melt the solid **(1)** or dissolve the solid in water. **(1)**

1.4 In both cases the ions are then free to move around. **(1)**

2.1 The silver liquid is (molten) sodium, the red-brown liquid is bromine. **(1)**

2.2 The colourless gas is hydrogen, orange solution is bromine. **(1)**

2.3 The lead ions have gained electrons. **(1)**

Page 96

Exam-style question

1.1 Aluminium is oxidised **(1)** because it gains oxygen **(1)**.

1.2 Heat it until it melts. **(1)**

1.3 $3e^-$ **(1)**; reduction **(1)**

O_2 **(1)**; oxidation **(1)**

Unit 6

Page 98

① **a** 9

b $2 \times 19 = 38$

② **a** sodium = 2 atoms

hydrogen = 1 atom

phosphorus = 1 atom

oxygen = 4 atoms

b relative mass = $(23 \times 2) + (1 \times 1) + (31 \times 1) + (16 \times 4) = 142$

③ **a** calcium = 1 atom

nitrogen = $(1 \times 2) = 2$ atoms

oxygen = $(3 \times 2) = 6$ atoms

b relative mass = $(40 \times 1) + (14 \times 2) + (16 \times 6) = 164$

Page 99

① relative atomic mass = $\left(35 \times \frac{75}{100}\right) + \left(37 \times \frac{25}{100}\right)$

$= 35.5$

② relative atomic mass = $\left(6 \times \frac{7.6}{100}\right) + \left(7 \times \frac{92.4}{100}\right)$

$= 6.924$

$= 6.9$ (to 1 decimal place)

③ relative atomic mass = $\left(28 \times \frac{92.2}{100}\right) + \left(29 \times \frac{4.7}{100}\right)$

$+ \left(30 \times \frac{3.1}{100}\right) = 28.109$

$= 28.1$ (to 1 decimal place)

④ **a** $79.9 \times 2 = 159.8$

b The relative atomic mass is closer to the atomic mass of bromine-79 than to the atomic mass of bromine-81, so bromine-79 must be more common.

Page 100

① one atom of Ca $\Rightarrow$ one unit of CaF_2

② $1\,g\ Ca \Rightarrow \frac{78}{40}\,g\ CaF_2$

$80\,g\ Ca \Rightarrow \frac{78}{40} \times 80\,g\ CaF_2 = 156\,g\ CaF_2$

③ **a** $Na_2CO_3 = (23 \times 2) + (12 \times 1) + (16 \times 3)$

$= 106$

b $NaCl = 23 + 35.5 = 58.5$

④ one unit of $Na_2CO_3 \Rightarrow$ 2 units of $NaCl$

$106\,g\ Na_2CO_3 \Rightarrow 2 \times 58.5 = 117\,g\ NaCl$

⑤ $1\,g\ Na_2CO_3 \Rightarrow \frac{117}{106}\,g\ NaCl$

$10.6\,g\ Na_2CO_3 \Rightarrow \frac{117}{106} \times 10.6\,g\ NaCl = 11.7\,g\ NaCl$

Page 101

① concentration $= \frac{36}{3} = 12\ g/dm^3$

② $400\,cm^3 = \frac{400}{1000} = 0.4\ dm^3$

concentration $= \frac{16}{0.4} = 40\ g/dm^3$

③ mass = 46.5 g

④ $250\ cm^3 = \frac{250}{1000} = 0.25\ dm^3$

mass = $1.60 \times 0.25 = 0.40\,g$

⑤ $500\ cm^3 = \frac{500}{1000} = 0.5\ dm^3$

mass = $32.4 \times 0.5 = 16.2\,g$

Page 102

① **a** 100

b $\left(10 \times \frac{19.9}{100}\right) + \left(11 \times \frac{80.1}{100}\right) = 10.801$

② 10.8

③ (32×4)

④ $2K \Rightarrow K_2SO_4$

$2 \times 39\,g = 78\,g\ K \Rightarrow 174\,g\ K_2SO_4$

⑤ $1\,g\ K \Rightarrow \frac{174}{78}\,g\ K_2SO_4$

$7.8\,g\ K \Rightarrow \frac{174}{78} \times 7.8\,g\ K_2SO_4 = 17.4\,g\ K_2SO_4$

Page 103

Exam-style questions

1.1 All three isotopes have the same number of electrons and protons (12 – the atomic number). **(1)** However, they have different numbers of neutrons **(1)**: magnesium-24 has 12 neutrons, magnesium-25 has 13 neutrons and magnesium-26 has 14 neutrons. **(1)**

1.2 $\left(24 \times \frac{79.0}{100}\right) + \left(25 \times \frac{10.0}{100}\right) + \left(26 \times \frac{11.0}{100}\right) = 24.32$

$= 24.3$ to 1 decimal place

numbers in correct formula **(1)**, *answer 24.32* **(1)**, *rounding to 24.3* **(1)**

2.1 $Mg + 2HCl \longrightarrow MgCl_2 + H_2$ **(1)**

2.2 The hydrogen is a gas so it escapes from the container. **(1)**

2.3 relative atomic mass of Mg = 24; relative formula mass of $MgCl_2$ = $24 + (35.5 \times 2) = 95$ **(1)**

$Mg \Rightarrow MgCl_2$ so $24\,g\ Mg \Rightarrow 95\,g\ MgCl_2$ **(1)**

mass of $MgCl_2 = \frac{95}{24} \times 12 = 47.5\,g$ **(1)**

3 concentration $= \frac{mass}{volume} = \frac{73.0}{2.00} = 36.5\ g/dm^3$

numbers in correct formula **(1)**, *answer 36.5* **(1)**

Page 104

Exam-style question

1 $\left(20 \times \frac{90.48}{100}\right) + \left(21 \times \frac{0.27}{100}\right) + \left(22 \times \frac{9.25}{100}\right)$

$= 20.1877$

$= 20.19$ (to 2 decimal places)

numbers in correct formula **(1)**, *20.1877* **(1)**, *rounding to 20.19* **(1)**

2 Relative formula mass of NaOH = 23 + 16 + 1 = 40

Relative formula mass of Na_2SO_4 = $(23 \times 2) + 32 + (16 \times 4)$ = 142 **(1)**

$2 \times NaOH \Rightarrow Na_2SO_4$ **(1)**

$2 \times 40\,g = 80\,g\ NaOH \Rightarrow 142\,g\ Na_2SO_4$

mass of Na_2SO_4 $= \dfrac{142}{80} \times 16$ **(1)** $= 28.4\,g$ **(1)**

3 concentration $= \dfrac{mass}{volume} = \dfrac{490}{100} = 49\ g/dm^3$

numbers in correct formula **(1)**, *answer 49* **(1)**

Unit 7

Page 106

(1)

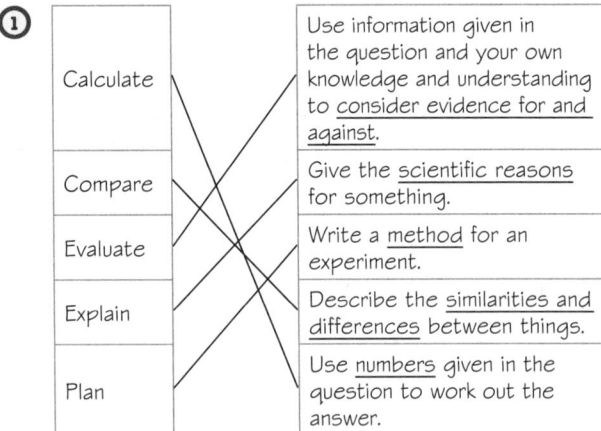

Calculate	Use information given in the question and your own knowledge and understanding to <u>consider evidence for and against</u>.
Compare	Give the <u>scientific reasons</u> for something.
Evaluate	Write a <u>method</u> for an experiment.
Explain	Describe the <u>similarities and differences</u> between things.
Plan	Use <u>numbers</u> given in the question to work out the answer.

(2) a Compare

b i protons (or electrons)

  ii electrons (or protons)

c neutrons

d All three isotopes have one proton in their nucleus and one electron shell containing one electron. However, each isotope has a different number of neutrons in its nucleus.

Page 107

(1) Explain

(2) structure, bonding

(3) ionic bonds, covalent bonds, intermolecular forces

(4) giant ionic lattice, giant covalent lattice, small molecules

(5) a metal; non-metal

b ionic lattice; strong

c non-metal

d intermolecular forces

e covalent lattice, strong covalent bonds

Page 108

(1) Evaluate

(2) a obtained in large amounts

b (obtained from) crude oil; carbon dioxide, sulfur dioxide, nitrogen oxides and particulates released

c produced from water; the only product is water

d produced by electrolysis; only a few places where a driver can get hydrogen; is a gas and has to be stored at high pressure in heavy tanks

(3) a There are large supplies of crude oil so petrol can be produced in large amounts. It is a liquid so it is easily stored and easily transferred to a car. There are many petrol fuel stations.

b Crude oil is non-renewable so supplies will run out. Carbon dioxide is a greenhouse gas. Sulfur dioxide and nitrogen oxides cause acid rain; particulates cause breathing problems.

c Hydrogen is produced from water which is in limitless supply. Burning hydrogen produces only water so causes no pollution.

d Hydrogen is produced by electrolysis. This needs electricity, so it is expensive and the generation of the electricity may involve burning fossil fuels which releases carbon dioxide. Hydrogen is a gas so it is difficult to store, and there are not many hydrogen fuel stations.

(4) Answers should cover all the points identified in the answers to (3), arranged in a sensible and logical order.

Page 109

(1) The table contains data about the energy used to manufacture bags (in MJ) and the waste mass formed (in g).

(2) positive aspect = better for the environment; negative aspect = is misleading

(3) energy consumption and waste mass

(4) using **just** this information

(5) More energy is used to manufacture a cotton bag than a single-use plastic bag. However, if the cotton bag is used two or more times, less energy is used overall.

The amount of waste mass produced in making the cotton bag is greater than that for the plastic bag, and this waste might have to be disposed of in landfill or by burning.

Compared to a single-use plastic bag, a cotton bag uses less energy and produces less waste if it is used more than four times.

Page 110

(1) a Describe; explain

b lithium; sodium; potassium

c describe the observations; explain how these observations can be used to place the metals in order of reactivity

(2) a The question does not ask about how the metal is prepared or added to the water. This part of the answer would gain no marks.

b This is a good, concise description with four observations. (i) It says where on the water the metal is (the surface). (ii) It states that the metal moves. (iii) It states that the metal fizzes. (iv) It states that the metal disappears. This description is also useful because it says how fast the reaction happens (it fizzes slowly).

c This is a good idea because it deals with all three metals, making sure that nothing is left out.

d A lilac flame is seen when potassium is dropped into a trough of water.

e (most reactive) potassium, sodium, lithium (least reactive)

Page 111

(1) Use a measuring cylinder to measure out some hydrogen peroxide solution and place it in a flask. Connect a gas syringe to the flask and measure the volume of oxygen produced every minute for 10 minutes.

Now repeat the experiment. Use the same volume of the same hydrogen peroxide solution, but add one spatula of P. Again, measure the volume of oxygen produced every minute for 10 minutes.

Repeat the experiment again adding Q rather than P. Then repeat it a final time, using R. Add the same amount of the substance (P, Q or R) each time. Keep everything else the same.

If one of the substances (P, Q or R) is a catalyst, the reaction will produce oxygen gas at a much faster rate when that substance is added to the hydrogen peroxide than when nothing is added.

Page 112

Exam-style question

1 Potable water is water that is fit to drink. It is made from fresh water from lakes / rivers / ground water by passing the water through filter beds and then sterilising the water with chlorine, ozone or UV light. Potable water can be made from sea water by distillation or reverse osmosis.

Making potable water from sea water is expensive because a large amount of energy is required. The United Kingdom does not use this method because it has good supplies of fresh water which can be made potable using the cheaper methods of filtering and sterilising.

Other countries which have limited supplies of fresh water must use the more expensive method of desalination to produce potable water from sea water.

Physics

Unit 1

Page 114

1 the dot 4

2 a weight; gravity is an acceptable answer, but the force of gravity on an object is called the object's weight

b the downwards arrow is longer

3 a the same

b accelerating

4 Force W and Force D must be equal because the speed is not changing

Page 115

1 a, b air; left (backwards)

water; up (and left is also an acceptable answer)

rope; right (forwards)

2 a gravity, magnetism

b gravity

3

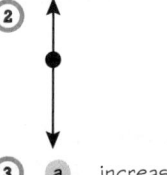

upthrust (water)

drag (air) ←——●——→ tension (rope)

weight (gravity)

4 magnetism: at a distance, upwards
weight (gravity): at a distance, downwards
tension: contact from the string, downwards
not air resistance: the clip is not moving through the air so there is no air resistance.

Page 116

1 stays the same

2 increases

3 a 5 N right + 20 N right = 25 N right

b 10 N right − 4 N left = 6 N right

4 a A 100 N right

B 300 N right

C 400 N left

b i A and B

ii C

5 stays the same

Page 117

1 ←——●——→ (arrows can be any length)

2 increases

3 a to the right

b i smaller

ii still increasing

iii still accelerating

4 a ←——●——→ (arrows must be equal length)

b zero

c stays the same

Page 118

1 gravity, air resistance

2 resultant force and acceleration

3 terminal velocity

4 The second answer, because it links the ideas about resultant force and acceleration to the situation in the question. Although the first answer talks about forces and changing speed, the important idea that resultant force causes acceleration is missing.

Page 119

1 the skydiver's weight or gravity, and air resistance

2

↑
|
●
|
↓

3 a increases

b stays the same

4 decreases

5 zero

6 decreases

7 zero

8 As the skydiver falls towards the ground they will reach a maximum speed (or terminal velocity) because the gravitational pull of the Earth is greater than the air resistance.

However, as the velocity of the skydiver increases the air resistance increases as well until the two forces are equal. This means that the skydiver will fall at a constant speed.

Page 120

Exam-style question

1.1 When the ball is dropped into the oil it accelerates because the gravitational pull of the Earth is greater than the resistance of the oil (upthrust). As the velocity of the ball increases so does the upthrust until the two forces are equal. This means the ball will not fall at a constant speed. The speed the ball reached before air resistance started to act upon it was the maximum speed (or terminal velocity).

Unit 2

Page 122

① neutrons, protons

② Atoms with the same number of protons but different numbers of neutrons

③

Radiation type	Description
alpha	electromagnetic radiation
beta	a high-speed electron
gamma	two protons and two neutrons, the same as a helium nucleus

④

Property	Description
penetrating	can easily remove electrons from atoms they collide with
ionising	can easily pass through materials

⑤ alpha ☐1☐   beta ☐2☐   gamma ☐3☐

⑥ Any two from: move away from it; put some shielding between you and the source; reduce the time you are near the source

Page 123

① radiation; source / radioactive material; surroundings

② rocks, nuclear power stations, smoke alarms

③ A worker at a factory accidentally breathing-in radioactive dust.

A hospital patient injected with a radioactive material.

④ 'so you don't get the radioactive source / material in you' OR 'so you are not contaminated by the radioactive source / material'

Page 124

① 400, two, 6 hours, 1 half life, ÷2; answer = 3 hours

② **a** 3 hours

b $\frac{14.3}{4}$ = 3.575 days

c 6 months

③

	3 days	+	3 days	+	3 days	+	3 days	= **12 days**
640	→		320	→	160	→	80	→	40

Answer = 40 kBq

④

	4 days	+	**4 days**	+	4 days	= **12 days**
1200	→		600	→	300	→	150

Answer = 150 Bq

Page 125

① **a**, **b** and **e**

②

	Source emitting alpha radiation	Source emitting beta radiation	Source emitting gamma radiation
Source on your clothes	☐	☐	✓
Source inside the body	✓	☐	☐

③ **a** and **c**

④ Gloves reduce the risk of getting the source on your hands **(1)**; a mask reduces the risk of breathing in particles of the source **(1)**.

Page 126

① **a** First line

b There should be four arrows: 800 → 400 → 200 → 100 → 50 so answer is 50 MBq

② **a** irradiated, ionising

b penetrating

c 'get radiation inside' should be 'get a radioactive source / radioactive material inside'

d 12 hours is two half-lives so the activity will be $\frac{1}{4}$ not zero. The student thinks the activity drops by the same amount every 6 hours, rather than by the same proportion (i.e. half) every 6 hours.

③ **a** The doctor is at risk because she might use technetium-99 every day and might get it on her clothes. The patient is at risk because the radiation will be in them for six hours and radiation causes cancer so they might die from it.

b This answer would gain no marks, because the student has not explained how to minimise the risks to doctors or patients from the radiation.

Page 127

Exam-style question

1.1 **i** 800 MBq, 6 hours

ii

	6	+	6	+	6	+	6	= 24 hours
800	→	400	→	200	→	100	→	50

Answer: 50 MBq

1.2 **i** patient

ii Use only a small amount of Tc-99

iii Standing close to the source; carrying / injecting the source

iv Any answer from: move away from the patient as soon as possible after injecting them; stand behind a screen; keep the source in a lead box; use a lead-lined syringe, etc.

(No marks for 'wear gloves / wear mask / etc.' because gamma radiation is very penetrating)

v Get Tc-99 on skin, clothes or in mouth etc.

vi Wear gloves, mask, etc.

vii (one mark for each of the following to a max of 4 marks)

Healthcare professionals need to monitor their exposure so that they do not pass the recommended annual dose of radiation – usually using a dosimeter.

Radioactive substances should only be used when absolutely necessary, and the specific circumstances of the patient must be considered, e.g. would not use radioactive substances on a woman who is pregnant or breastfeeding.

Radioactive substances must be stored correctly to reduce the risk of contamination.

Healthcare professionals should use appropriate personal protective equipment (PPE) when handling radioactive substances – e.g. a shielded syringe made of Tungsten can be used when administering Technetium-99.

Patients must listen to and follow all instruction from the healthcare professionals to reduce the chances of having to repeat the procedure which would expose them to more radiation.

Patient may be given personal protective equipment to protect areas of the body, e.g. a lead apron during an x-ray to protect areas of the body that do not need to be imaged.

Page 128

Exam-style question

1.1 Radiation is <u>ionising</u> (1). This can <u>damage cells</u> (1), increasing the risk of cancers in the future. Humans may become contaminated / it is not safe or ethical for humans to be irradiated (1). It is safe for robots because <u>they are not made of cells</u> / they are made of metal (1).

Unit 3

Page 130

① a gas
   b liquid
   c solid

②
Solid	Rapid, random motion
Liquid	Vibrate around fixed positions
Gas	Move past one another

③
Melting	Turning straight from solid into gas
Boiling	Turning from liquid into gas
Sublimation	Turning from solid into liquid

④ freezing, condensing

⑤ a kinetic energy
   b density

Page 131

① a temperature
   b expands
   c density
   d volume

②
Property	Explanation
A gas expands to fill its container because	particles vibrate around fixed positions.
A solid cannot flow because	particles are packed closely together.
A liquid has a high density because	there are no forces between the particles.

③ a a high temperature
   b The second answer is better because it is clearer. The first answer says the particles expand: particles stay the same size but the bar expands. The first answer says 'it melts' but it isn't clear whether 'it' means a particle or the bar. The second answer states clearly that the bar melts.

Page 132

① internal; kinetic / potential; potential / kinetic

② potential energy; temperature, kinetic energy

③ a 'Melt' marked on the first / lower flat section of the graph
   b 'KE increase' marked on the three upward sloping parts of the graph

④ a increases
   b stays the same
   c kinetic energy

⑤ When water is heated, the water particles gain kinetic energy. Temperature only depends on particle kinetic energy so the temperature increases.

⑥ When a substance changes state, the particles gain potential energy. Temperature depends on particle kinetic energy (not potential energy) so the temperature stays the same / constant when changing state.

Page 133

① rapid, random motion

② a gas
   b particles

③ particles (of the gas) colliding with the walls (of the container)

④ speed of particles; mass of particles; number of particles; volume of container

⑤ speed of particles

⑥ number of particles

Exam-style question

1.1 The pump increases the number of air particles in the tyre. Pressure is caused by particles colliding with the walls. There will be more collisions if there are more particles so the pressure increases.

Page 134

① Replace 'Hotter particles' with 'Particles with more kinetic energy'. Replace 'hit' with 'collide with'.

② pressure increases / collide more often

③ heating increases the particles' kinetic energy so they are moving faster.

④ There are more collisions (per second) with the walls (because there are now more particles to collide with them).

Page 135

1. Particles (of the gas) colliding with the walls (of the container)

2. The particles gain kinetic energy so they move faster

3. Increases

4. When the particles move faster they collide with the container walls more often / harder, so the pressure increases.

5. **Exam-style question**

   Pressure is caused by collisions between gas particles and the container walls. Heating the gas increases the particles' kinetic energy so they are moving faster and collide with the walls more often and with more force. This causes the pressure to increase.

   Marks for: increases pressure (1); then any two of: pressure is due to collision with walls (1), heating causes particles' KE to increase (1), collisions occur with greater (average) force (when KE is increased) (1), collisions occur with greater frequency / more often (when KE is increased) (1).

6. Both the mass of the gas and volume of the container would affect the pressure if they changed.

Page 136

Exam-style question

1.1 If the temperature of the gas increases, the pressure inside the can will increase (1). If the pressure gets too high, the can could explode (1). This could injure people (1).

Exam-style question

2.1 The clothes dry by evaporation (1). The rate of evaporation will be quicker if the surface area of the clothes is greater (1). Spreading out the clothes will increase their surface area (1).

Unit 4

Page 138

1.

W	Current	Flow of charge
C	Potential difference	The electrons that flow in the wire
V	Charge	Same as voltage
A	Power	Energy transferred in a given time

2. **Current** can flow through the circuit easier when there's less resistance.

   A normal light bulb uses more **energy** each day than an LED bulb.

2. a i A   ii V   iii <omega, Ω>

   b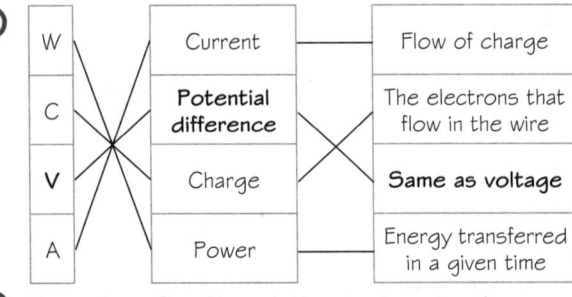

4.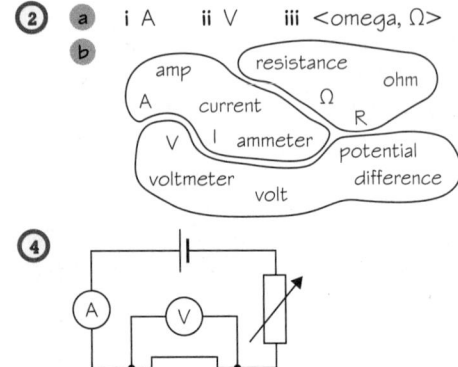

5. variable resistor

6. controls the current by changing the resistance

7. $R = \dfrac{V}{I} = \dfrac{6}{0.1} = 60\,\Omega$

Page 139

1. grey (+) circle = copper atom; small red dot = free electron

2. negative

3. arrow pointing clockwise, from negative to positive

4.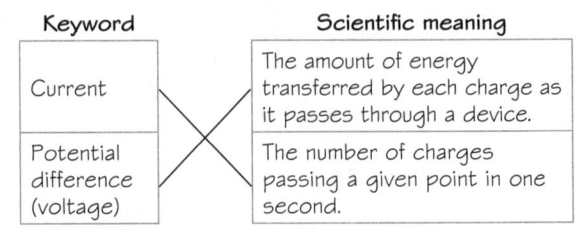

Keyword	Scientific meaning
Current	The amount of energy transferred by each charge as it passes through a device.
Potential difference (voltage)	The number of charges passing a given point in one second.

5. $I = \dfrac{Q}{t} = \dfrac{3\,C}{60\,s} = 0.05\,A$

6. $V = \dfrac{E}{Q}$ so $E = V \times Q$ so $E = 5 \times 9 = 45\,J$

Page 140

1. flow; charge (or electrons)

2. a Current will decrease

   b Adding more devices increases the total resistance and $I = \dfrac{V}{R}$

3. The reading will decrease, because adding a second bulb increases the resistance (in the circuit).

4. toaster, hairdryer, bulb

5. The device might catch fire.

6. a $\dfrac{4\,V}{0.2\,A} = 20\,\Omega$

   b $\dfrac{12\,V}{0.2\,A} = 60\,\Omega$

Page 141

1. Battery = 1.5 V, Mains = 230 V

2. brown; green/yellow; blue

3. live

4. increases; hot; melts (not blows / snaps); breaks

5. earth wire, fuse

6. washing machine, iron, lawnmower

Page 142

1. decreases (a lot)

2. increases

3. melts

4. When there is a fault the current ~~goes up~~ **increases** and the fuse ~~blows~~ **melts** making the appliance safe.

5. There is no longer a complete circuit so no current can flow.

6. earth wire and fuse

7.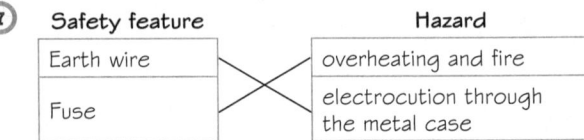

Safety feature	Hazard
Earth wire	overheating and fire
Fuse	electrocution through the metal case

Page 143

1. increases

2. The thin wire gets very hot and melts.

③ There is no longer a complete circuit, so current can flow.

④ If a fault occurs the current will increase **(1)**. This will make the fuse wire get very hot and melt **(1)**. This breaks the circuit, so no current can flow, making the washing machine safe **(1)**.

Page 144

Exam-style question

1.1 A thinner wire will heat up more than a normal / thicker wire **(1)**; so it may overheat / catch fire **(1)**.

Unit 5

Page 146

① 10 m/s

② m (metres)

③ s (seconds)

④ Speed = $\dfrac{100}{12}$ = 8.3 m/s

⑤ amps (A) (or ampere)

⑥ volts (V)

⑦ pd = 2 A × 100 Ω
= 200 V (volts)

Page 147

① ⓐ braking force

ⓑ distance and work done

ⓒ work done = force × distance

ⓓ braking force = work done ÷ distance = 2000 N ÷ 20 m = 100 N/m

② Before: momentum = mass × velocity = 0

After: momentum = mass × velocity = 0.1 kg × 20 m/s = 2

Change in momentum = 2 − 0 = 2 kg m/s

③ Before: kinetic energy = $\frac{1}{2}$ × mass × speed²
= 0.5 × 0.05 × 35²
= 30.625 J

After: kinetic energy = $\frac{1}{2}$ × mass × speed²
= 0.5 × 0.05 × 25²
= 15.625 J

Change in kinetic energy = 30.625 − 15.625 = 15 J

Page 148

① m = 0.001, G = 1 000 000 000, M = 1 000 000, k = 1000

② 15 kΩ = 15 000 Ω
10 mA = 0.010 A (i.e. 0.01)
pd = current × resistance = 0.01 × 15 000 = 150 V

③ ⓐ 50 g = 0.05 kg

ⓑ weight = mass in kg × 10 N/kg = 0.5 N

④ ⓐ 1 minute = 60 s

ⓑ charge = current × time = 5 × 60 = 300 C

⑤ ⓐ The total mass is 20 kg + 60 kg = 80 kg

ⓑ kinetic energy = $\frac{1}{2}$ × mass × speed²
= $\frac{1}{2}$ × 80 × 10² = 4000 J

Page 149

① 10; 5

② momentum = mass × velocity
30 = mass × 6
divide both sides by 6: mass = 5 kg

③ Energy = J, Mass = kg, Time = s, Charge = C, Pd = V, Power = W

④ Student has used a lower case j – joules must be with a capital J.

⑤ ⓐ 12 = 2 s.f.

ⓑ 2.7 = 2 s.f.

ⓒ 1.06 = 3 s.f.

ⓓ 2.0 = 2 s.f.

Page 150

① ⓐ They cannot use this equation because they don't know the change in speed.

ⓑ The other mistakes they have made are: (1) time has to be in seconds so they should have used 60 (seconds) rather than 1 (minute); (2) they have used the wrong unit (m/s rather than m/s²).

② ⓐ In m/s² the square is part of the unit: it does not mean you have to square the answer.

ⓑ The force should be the resultant force: 1800 − 950 = 850 N

The mass should be the total mass: 900 + 150 = 1050 kg

Page 151

① acceleration = $\dfrac{\text{change in speed}}{\text{time taken}}$

force = mass × acceleration

② force, mass

③ overall force, i.e. forward force − drag force = 1800 − 950 = 850 N

④ total mass of car + passengers = 900 + 150 = 1050 kg

⑤ m/s²

⑥ F = ma; so a = $\dfrac{F}{m}$ = $\dfrac{850\,\text{N}}{1050\,\text{kg}}$ = 0.81 m/s² to 2 s.f.

Page 152

Exam-style questions

1.1 total energy transferred = 200 + 100 = 300 J **(1)**

power = $\dfrac{\text{energy transferred}}{\text{time taken}}$ **(1)**

= $\dfrac{300}{1\ \text{minute}}$ **(1)**

= $\dfrac{300}{60}$ **(1)**

power = 5 W (or watts)

2.1 wave speed (m/s) = frequency (Hz) × wavelength (m) **(1)**

frequency = speed ÷ wavelength
= 3.0 × 10⁸ ÷ 20 **(1)** = 1.5 × 10⁷ Hz **(1)**

Unit 6

Page 154

① ⓐ Energy transferred in J; Time in s

ⓑ unit(s)

② ⓐ 300

ⓑ 100 J

③ 10, 15

④, ⑤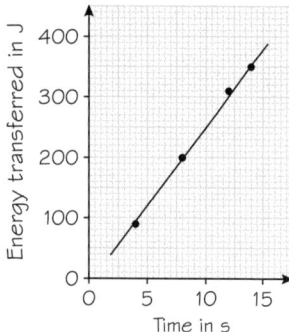

Page 155

① 10

② 1

③ 10

④ 2

⑤ length value = 10 + 6 = 16 cm

force value = 10 + 4 = 14 N

Page 156

① A and C

②
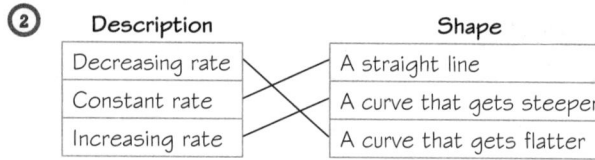

Description		Shape
Decreasing rate		A straight line
Constant rate		A curve that gets steeper
Increasing rate		A curve that gets flatter

③ A

④
  a B: As the force increases, the speed decreases at a decreasing rate.

  b C: As the force increases, the speed increases at a decreasing rate.

  c E: As the force increases, the speed decreases at a constant rate.

  d F: As the force increases, the speed decreases at an increasing rate.

Page 157

① Gradient

$$= \frac{\text{vertical distance between the start and end points}}{\text{horizontal distance between the same two points}}$$

$$= \frac{80\,\text{m} - 0\,\text{m}}{100\,\text{s} - 0\,\text{s}}$$

$$= \frac{80\,\text{m}}{100\,\text{s}}$$

$$= 0.8\,\text{m/s}$$

② Student A has just used the last point on the graph (60 m/s after 100 s) instead of looking at the section from 40 s to 100 s. They should have used $\frac{60 - 40}{100 - 40}$ as their calculation.

Student B has looked at the section from 60 s to 80 s. This would work but they have mis-read the scale on the speed axis – their 47 – 43 m/s should be about 54 – 46 m/s.

③ The correct gradient using Student A's method is

$$\frac{60 - 40}{100 - 40} = \frac{20}{60} = 0.33\,\text{m/s}^2$$

The correct gradient using Student B's method is

$$= \frac{54 - 46}{80 - 60} = \frac{8}{20} = 0.4\,\text{m/s}^2$$

Student B wasn't able to read the scale exactly so has an inaccurate answer; it is better to use the biggest range you can when finding the gradient, like Student A tried to do.

Page 158

① It then increases at a decreasing rate.

② **a** 0.5

  b They counted each small square above 10 as 1 N, so they got 10 N + 2 squares = 12 N (whereas 12 N is 10 N + 4 squares)..

Page 159

1.1 extension increases, straight line, until 15 N, decreasing gradient

1.2 **i** horizontal line drawn on graph, across from 12 on the y-axis

  ii 4.4 cm

1.3 **i** cm

  ii N/m

  iii 6 cm = 0.06 m

  iv 5 N and 0.02 m

Page 160

Exam-style question

1.1 As the angle of refraction increases, the angle of incidence increases at an increasing rate.

1.2 26 degrees

Unit 7

Page 162

① Compare

② light waves, sound waves

③ whereas, faster

④ **a** Student B's answer is better.

  b Repeat / explanation and formula for finding volume / formula for finding density

⑤ Any relevant detail, e.g. formula for calculating density

⑥ The section about how to find the volume by the displacement method – Student B should have described the method.

Page 163

① Describe the methods you would use to determine the effectiveness of an insulator.

Plan an experiment to investigate how the force on a spring affects its extension.

Explain how you would measure the resistance of a thermistor at different temperatures.

② **a**
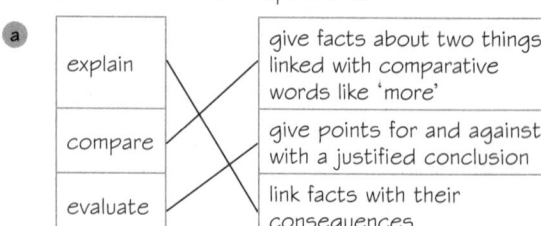

explain		give facts about two things linked with comparative words like 'more'
compare		give points for and against with a justified conclusion
evaluate		link facts with their consequences

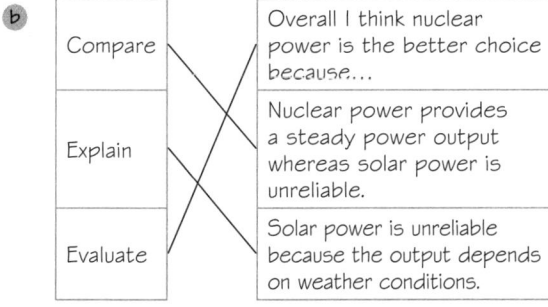

b

Compare	Overall I think nuclear power is the better choice because...
Explain	Nuclear power provides a steady power output whereas solar power is unreliable.
Evaluate	Solar power is unreliable because the output depends on weather conditions.

(Compare links to Solar power is unreliable; Explain links to Overall I think nuclear power; Evaluate links to Nuclear power provides)

② ideas about transverse and longitudinal waves

Page 164

① 1. Measure the length of a trough

2. Half-fill the trough with water

3. Raise one end a little and drop it

4. At the same time, start the stopwatch

5. Stop the stopwatch when the wave has travelled to the end of the trough and back again

6. Repeat the previous steps two more times

7. Calculate the mean time taken

8. Use the equation: $\text{speed} = \dfrac{\text{length of trough} \times 2}{\text{mean time}}$ to find the speed of the waves

② **a** Therefore, As a result

b 1. A thermistor has a lower resistance when the temperature is high.

2. Therefore, when the temperature in the boiler is too low

3. the resistance of the thermistor will be high.

4. As a result, the current in the circuit will increase.

5. This can be detected and used to turn the boiler on.

Page 165

① **a** circular / shaped like a bar magnet; depends on distance / uniform;

depends on current / depends on current; weaker / stronger

b straight, iron core, electromagnet, solenoid

② **a** Compare

b the factors that affect the strength of the magnetic fields

③ **a** around an HB pencil

b Make sure you have six wires, three red and three black.

c Go to the bench with the meters

Page 166

① **a** No

b They have included lots of irrelevant details such as safety advice – bags out of the way, hair tied back – this was not asked for in the question.

c ~~First go and collect all your equiptment and get it all together on the table so that it is easy to set up. Make sure your bags and folders are out of the way and if you are a girl you have tied back your hair. Then go to the tap to~~ fill up the measuring cylinder with water. ~~You should use~~ 100 ml ~~but it doesn't matter if it is more or less but you must~~ write down how much is in it ~~so you know how much you have got. Then~~ put the stone in the water ~~and~~ see how much it goes up.

Write down the volume of the stone then divide it by the mass and that's the density.

② No – they should explain that the volume of the stone is the same as the volume of water displaced (the amount by which the reading from the measuring cylinder goes up)

③ 'equiptment' – this should be 'equipment' or, better, 'apparatus'

Page 167

① Describe

② an investigation to find the density of the stone

③ any causes of inaccuracy in the measurements

④ **a** mass and volume (of stone)

b $\text{density} = \dfrac{\text{mass}}{\text{volume}}$

⑤ Zero a balance measuring up to 1 kg in g to 1 d.p. Place the stone on the balance and record the reading.

⑥ Half fill a large measuring cylinder with water and record the reading on the scale in cm³. Carefully lower the stone into the water, being careful not to splash, and record the new reading in cm³. Subtract the first reading from the second and record this as the volume of the stone.

⑦ **a** volume

b Accept any suitable answer, e.g. some water may splash out of the cylinder, harder to read / reference to meniscus, cylinder has lower resolution than balance, some of the stone may fall off / dissolve

⑧ Sample answer:

• Zero a balance measuring in g up to 1 kg to an accuracy of 1 d.p. Place the stone on the balance and record the reading.

• Half-fill a measuring cylinder with water and record the reading on the scale in cm³.

• Carefully place the stone in the cylinder, being careful not to splash, and record the new reading in cm³. Subtract the first reading from the second and record the difference as the volume of the stone.

• Calculate the density of the stone using the formula: $\text{density} = \dfrac{\text{mass}}{\text{volume}}$

• This answer may not be completely accurate because the measurement of the volume may have been inaccurate:

  • Some water may have splashed out of the measuring cylinder when the stone was put in.

  • The readings from the measuring cylinder may have been inaccurate, because of the meniscus.

  • The scale on the measuring cylinder has a lower resolution than the balance.

Page 168

Exam-style question

Sample answer:

• Zero a balance measuring in g up to 1 kg to an accuracy of 1 d.p.

• Measure and record the mass of the empty cup using the balance.

• Measure and record the mass of the cup + water on the balance.

• Subtract the first mass from the second to find the mass of water in the cup.

• Connect a joulemeter to the immersion heater.

- Use a thermometer accurate to 0.5°C to record the temperature at the start.
- Turn on the heater.
- Wait for the temperature to rise by at least 10°C.
- Record the temperature and the reading on the joulemeter.
- Subtract the final temperature from the initial temperature to find the temperature change.
- Calculate the specific heat capacity using the formula:

specific heat capacity =

$$\frac{\text{joulemeter reading}}{\text{mass of water} \times \text{temperature change}}$$

- Uncertainty / inaccuracy could come from e.g. spilled / evaporated water, energy transfer to the surroundings.